KENTUCKY FARMER INVENTS WIRELESS TELEPHONE!

But Was It Radio?

Facts and Folklore About Nathan Stubblefield

BOB LOCHTE

All About Wireless, Murray, Ky.

Kentucky Farmer Invents Wireless Telephone!
But Was It Radio?
Facts and Folklore About Nathan Stubblefield

All About Wireless
PO Box 1194
Murray KY 42071

Publishers Cataloging-in-Publication Data

Lochte, Robert H. (1950-)
Kentucky farmer invents wireless telephone! : but was it radio? : facts and folklore about Nathan Stubblefield / Bob Lochte.
p. cm.
Includes bibliographical references.
LCCN 2001012345
ISBN 0-9712511-9-3
1. Stubblefield, Nathan Beverly, 1860-1928. I. Title.
2. Telephone, Wireless—Kentucky—History.
3. Radio—Kentucky—History.

TK6545.S82L63 2001
384.54/0976/992

Printed by Innovative Printing and Graphics,
Murray KY

To
Larry Albert

He showed me
how it worked.

ACKNOWLEDGEMENTS

So many people have helped me on this project that it's hard to thank them all. Here at Murray state University, I've been fortunate to have access to the Stubblefield archives at the Pogue Library. Dortha Bailey, Ernie Bailey, Dean Coy Harmon, and Keith Heim have been wonderful. At the Wrather Museum, Kate Reeves and Sally Alexander opened their Stubbelfield collection for me.

Three Deans, Gary Hunt, Ted Wendt, and Dannie Harrison have generously supported my research. And two chairs, Bob McGaughey and Jeanne Scafella have provided me time to work on this and other manuscripts. Orville Herndon made the computers work for me, and my colleagues John Dillon, Joe Hedges, Bill Mulligan, Peter Murphy, Bob Valentine, Allen White, and Ken Wolf reviewed the manuscript at various phases and offered sincere and practical criticism. And where would I be without Al Chan and his web page wizardry?

Several students worked on this project with me. Joe Kayse designed the cover artwork. Troy Fields showed me some Photoshop tricks. And Tory Holton handled all my publicity and marketing efforts.

Research often involves travel, and that costs money. I'm most grateful for the grants that I received from Peter Whaley and the CISR here at Murray State and from the Southern Regional Education Board in Atlanta. I offer a very special thanks to Elliot Sivowitch at the Smithsonian for helping me get a fellowship to work with him at the National Museum of American History.

Finally, thanks to my wife, Kate Lochte, for putting up with this project over the years.

Bob Lochte, Ph.D.

Murray, Kentucky
July 2001

Kentucky Farmer Invents Wireless Telephone!

Table of Contents

Kentucky All Over by Edwin Finch

RADIO WAS INVENTED BY A KENTUCKIAN.

NATHAN B. STUBBLEFIELD, OF MURRAY, CALLOWAY COUNTY, AN ECCENTRIC ELECTRICIAN, SUCCESSFULLY DEMONSTRATED HIS INVENTION, THE WIRELESS TELEPHONE, WHICH HE CALLED "RAIDIO," ON THE PUBLIC SQUARE IN MURRAY IN 1892, THREE YEARS BEFORE MARCONI SENT DOTS AND DASHES THROUGH THE AIR.

ON MARCH 20, 1902, STUBBLEFIELD TALKED FROM THE LAUNCH "BARTHOLDI," IN THE POTOMAC, AT WASHINGTON, TO SCIENTISTS ON THE RIVERBANK THROUGH HIS WIRELESS TELEPHONE, AND THE NEWSPAPERS ACCLAIMED THIS FEAT AS SOMETHING WHICH "IT IS BELIEVED WILL MARK AN IMPORTANT DEVELOPMENT IN THE TRANSMISSION OF SPEECH."

THE PATENTS OBTAINED ON HIS INVENTION INDICATE THAT HE HAD DISCOVERED FUNDAMENTAL PRINCIPLES OF BROADCASTING.

A GROUP OF NEW YORKERS FORMED A STOCK COMPANY CALLED THE "WIRELESS TELEPHONE COMPANY OF AMERICA" TO PROMOTE STUBBLEFIELD'S INVENTION, BUT FOR SOME UNKNOWN REASON HE WOULD NOT PART WITH HIS SECRETS. HE WAS SAID TO HAVE REFUSED AN OFFER OF $500,000 BECAUSE HE THOUGHT HIS INVENTION WAS WORTH TWICE THAT MUCH.

HE CAME BACK TO CALLOWAY COUNTY AND LIVED THE LIFE OF A RECLUSE IN A CABIN BACK IN THE HILLS, WHERE HE CONTINUED HIS EXPERIMENTS. HIS NEIGHBORS TOLD WEIRD TALES OF SEEING LIGHTS APPEAR IN TREES AND ALONG FENCES, AND OF VOICES COMING OUT OF THE AIR.

IN MARCH, 1928, STUBBLEFIELD CAME TO THE HOME OF MRS. L. E. OWEN AND ASKED HER TO WRITE HIS BIOGRAPHY, SAYING THAT HE HAD PERFECTED A LIGHT FROM ELECTRICITY TAKEN OUT OF THE AIR. TWO WEEKS LATER, MARCH 28, HE WAS FOUND DEAD IN HIS LONELY CABIN.

Louisville Courier Journal, 1946

The Old Wizard astounds big-city financiers and startles an incredulous hillbilly with his amazing inventions.

How It Began

"What about Nathan Stubblefield? Didn't he invent radio?"

An inquisitive eleven-year-old blurted out the questions, and I was momentarily flustered. Here I was, a college professor, former radio station owner, an "expert" on broadcasting, and I couldn't come up with an answer for the curious middle-schooler. I made a mental note to find out more about Murray, Kentucky's local hero. A few months later, I was doing research on the early history of radio and kept stumbling across references to Stubblefield, something about an 1892 broadcast. That was 1990. Two years later, I was recreating that century old event and deep into Stubblefield lore. Little did I know that what should have been a celebration of Nathan's achievements would devolve into threats of lawsuits, the financial ruin of Murray's oldest broadcasting facility, and, momentarily, the end of a decades-old attempt to claim the birthright of radio. But confusion is not unusual where Nathan Stubblefield is concerned.

On North 4th Street in Murray, Kentucky, stands a Kentucky historical marker. Its confusing inscription reads:

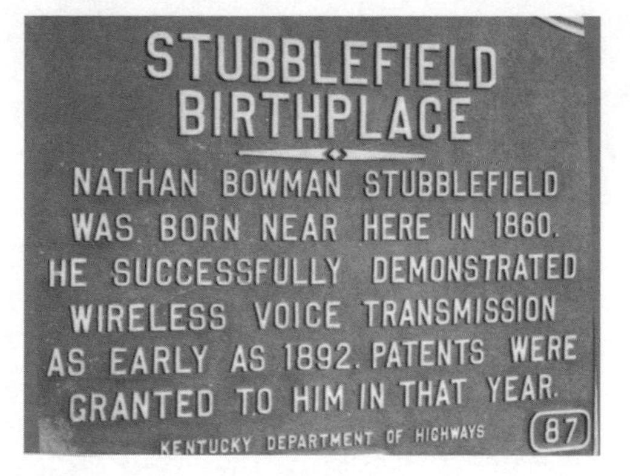

Almost every statement on that plaque is incorrect. The Kentucky Department of Highways got his name wrong.

Nathan didn't get any patents in 1892; no one knows where he was born, but the Stubblefield family farm was miles away, on the other side of town. The marker ought to commemorate his gravesite — it's just a few hundred yards north, in an old family cemetery.

There, Nathan's tombstone stands out because it's the newest one, installed in 1952, almost a quarter century after he died. It reads:

> Here lie the earthly remains of Nathan B. Stubblefield
> Nov. 22, 1860 - Mar. 28, 1928
> Inventor of Wireless Telephony, or Radio

At least the first three statements are true, probably. The bit about radio, though, has been the topic of discussion, confusion, and folklore around Murray, Kentucky for more than 70 years.

Bowman family cemetary near Murray
where Nathan Stubblefield is buried

Looking south from Nathan's grave, you can see the tip of a radio tower that belongs to WNBS-AM, whose call letters are Nathan's initials. The businessmen who put it on the air in 1948 believed that he invented radio, or at least thought the slogan "Murray, Kentucky, the Birthplace of Radio" was a catchy addition to the station identification. Other people, mostly outsiders, aren't convinced.

Nathan Beverly (the right middle name) Stubblefield was Calloway County, Kentucky's most famous resident. For the first half of 1902, newspapers from St. Louis to the East Coast extolled his exploits and scientific achievements. In the new field of wireless electric communication technology, some considered him Marconi's peer. But Nathan's notoriety was brief. Forgotten and alone, he came to a pathetic end twenty-six years later. As evidence of his past glory, however, his obituary appeared in the *New York Times* in 1928.

Then something remarkable happened. A young college professor and a group of eager journalism students investigated the details of Nathan's life. Their efforts forced the community to revisit the story and to commemorate Stubblefield's moment of glory. Remembrance led to embellishment, and embellishment to legend. Soon Nathan B. Stubblefield, the wireless inventor, became Nathan B. Stubblefield, the folk hero. By the present day, the images associated with both states of being are permanently entwined as a part of the local culture. Over the years, attempts to elaborate, enhance, and dramatize Nathan's saga have created a large body of misinformation and fantasy. The fiction gets repeated, sometimes for dubious purposes, as often as the truth does.

The first part of this book is Nathan's story. Several competent writers have told this before, so my effort here was largely to organize and summarize their work, adding my own interpretation where necessary. Here, the middle section, from about 1885 through 1912, is more detailed because there is more evidence of what really happened. Nathan's early and late years are not as clearly documented. The last decade is especially vexing. By then, Nathan was a peculiar old hermit and the subject of much speculation and gossip, making it difficult to get at the facts.

The second part is a chronicle of Nathan's legend. It commences within a few months of his 1928 burial in an unmarked grave. Along the way, Nathan has been the subject of newspaper features and magazine articles, nationwide radio broadcasts, a radio drama, a stage musical and a children's book, publicity campaigns, two doctoral dissertations, a television documentary, a traveling history show, museum exhibit, and dozens of World Wide Web pages. His story appears in books on Kentucky history and in chronicles of the radio industry. Journalists, scholars, and crackpots alike have told of his accomplishments. There are even some people who believe that Nathan cracked the code of telluric and cosmic forces by inserting a magic device near the roots of charmed oak trees. Who needs Harry Potter when we have such fascinating folklore right here?

It's a great yarn, and it begins with the first wave of settlers in Far West Kentucky during the early 19th century.

Nathan's Early Life

Surrounded by water, cut off from the rest of Kentucky, and occupied by Chickasaws, the western part of the Commonwealth was largely unsettled until the third decade of the 19th century. The pressure of western expansion and the demands of Revolutionary War veterans to claim the tracts of land granted to them by cash-strapped Virginia for their war service forced Kentucky to act. In 1818, former governor Isaac Shelby and General Andrew Jackson negotiated a treaty to purchase the territory from the Indians for $300,000. The US Senate ratified the treaty the following year and Kentucky acquired this parcel of land known as the Jackson Purchase.

Seven years later, Beverly Stubblefield moved west from North Carolina to claim his 5,166-acre land grant, among the largest tracts that belonged to the first wave of settlers. He made his home in New Concord, a few miles from Murray. In 1830, his wife bore him a son whom they named William "Billy " Jefferson.

Billy Stubblefield began a career as a farmer, but he lacked both interest and stamina for that vocation. He decided to go to law school in Louisville, then returned to set up practice in Murray, married Victoria Bowman, and began to raise a family. They had four sons, the third one named Nathan Beverly.

Young Nathan's life was confusing from the get go. Historians disagree on whether he was born in 1859 or 1860. The Census of 1860, however, only includes two sons in Billy and Victoria's household, whereas the 1870 Census lists Nathan, age 9. So it appears that 1860, probably in late November, is correct.

While Nathan was an infant, his father went off to war as a member of the Seventh Kentucky Regiment of the Confederate Army. Although Kentucky remained in the Union, the sentiments in the Jackson Purchase were with the South. Some of the Stubblefields were probably slave owners. The regiment elected Billy Stubblefield a captain, and he kept the nickname Captain Billy for the rest of his life. At the battle of Shiloh in 1862, he came down with pneumonia and had to return home. In 1869, his wife died, and a few years later Captain Billy remarried Clara Janes, the former governess to the children. Billy contracted pneumonia again in 1874 and died at the age of 44, leaving Nathan Stubblefield an orphan at age 14.

Nathan's immediate and extended families were financially secure, "passing rich" in the local vernacular. They had plenty of good farmland, and Captain Billy amassed a sizeable estate, for the era, from his law practice. Nathan enjoyed the relative luxury of private tutoring from a governess in his younger years and later attended the Male and Female Institute, a boarding school in nearby Farmington, until his father's death. It was about the best basic education available locally at that time. There were no high schools in the area. Originally, he intended to follow his father into the law practice but lost interest in that plan during his teenage years.

During the early years of his orphanage, Nathan developed three lifelong interests — farming, experimentation, and independence. He and a friend, Duncan Holt, made regular visits to the *Calloway County Times* office to read the periodicals. At some point, Nathan's interest drifted from *Practical Farmer* to *Scientific American*. Nevertheless, farming would be his primary activity for a few years.

In 1881, Nathan married Ada Mae Buchannan, a girl from Paducah whom he had known at the Farmington school. She was 17 and he 21. They moved into a 2-room frame house on an acre of his father's property about a mile west of Murray. Turning the plot into an extensive garden, Nathan produced just about every edible fruit and vegetable available. From this activity, he was able to feed the two of them and sell the excess every Saturday on the Murray square. To attract attention, he always wore his best suit and a Derby hat to market.

Soon, Nathan's family began to grow. In 1885, they had their first child, Carrie, who died in infancy, followed by a son, Bernard Bowman, in 1887. Frederic came a year later, but he too did not survive. Then there were two daughters, Patti Lee and Victoria Edison, in 1890 and 1892, two more sons, Nathan Franklin and Oliver Jefferson, in 1895 and 1897, and another daughter, Helen Gould, in 1901. Their last child, William Tesla, was born in 1903, but only lived 14 months.

As the family grew, Nathan expanded his enterprise to include farming a portion of his late father's land adjacent to his own. But the income from farming was never quite enough to support the burgeoning family, so Nathan turned his urge for experimentation in another direction — invention. He devised a convenient tool to light coal oil lamps without having to remove the glass chimney, got local attorneys to help him file a patent application, and a few months later, on November 3, 1885, he received US Patent no. 329,864 for the lighting device. What could be easier? It didn't matter that he never sold any of these gadgets. Nathan was convinced that his destiny and fortune lay in this direction.

For several years, Nathan had been interested in electrical devices. He had taught himself about circuits and batteries, and learned basic telegraphy. He added the new publication *Electrical World* to his reading list along with his old standby *Scientific American.* He started a small library of technical manuals and set up a 2'x2' shack behind the house as workshop. Eventually his expenditures on reading materials and supplies for his experiments squeezed the household budget severely.

Deprivation and poverty became his family's routine. After a while, they ceased to leave the farm. He didn't allow the children to go to public school or church or to have playmates at home. Partly this was due to Nathan's developing obsession with keeping his inventions a secret, but the embarrassment of not being able to afford proper clothes was another reason for the encroaching isolation. He kept a shotgun by the door to chase off intruders, and rigged the fences and gates with homemade alarms. In his proud independence, Nathan told anyone who offered to help that his family could get by on its own.

His relatives were aware of Nathan's eccentricities. When Nathan's older brother William died in 1894, he left an odd bequest that ultimately determined the disposition of the farm and Nathan's fate. In the will, William gave $750 jointly to Nathan's three children, Bernard, Pattie, and Victoria, who were then alive and directed his stepmother Clara to use the money to buy part of Captain Billy's farm, with the deed in the children's names. She was to do this "under the direction and with the consent of Nathan," but it was obvious that William wanted to assure that the children and not his brother benefited from this arrangement. He also left $50 for Nathan to use to buy seed, implements, and supplies needed to cultivate the land. Clara followed the directions and bought the property that Nathan was already using. He added to it his original plot for a total of 85 acres and used at least part of the money to plant a large apple and peach orchard.

Although there are scant details about this period of Nathan's life and the exact chronology is unclear, some interesting personality characteristics are apparent. He was a private man who nevertheless wanted public attention. He was sharp enough to learn about promotion, electricity, and patent law on his own. He wanted a better life for his family, perhaps out of personal pride at being born a Stubblefield. But he wanted to achieve success on his own. He saw electrical inventions as the route to prosperity, and named three of his children for famous electricians.

For Nathan, the nascent telephone business offered the most attractive opportunity to pursue his dreams.

Stubblefield's Telephone Business

By the mid-1880s, telephone service was diffusing rapidly throughout the cities of the United States. As many urban franchises were already granted, entrepreneurs were beginning to focus attention on smaller communities. A common challenge to all telephone ventures was the near monopoly that the Bell system had on important patents for the electrical telephone. To set up a local telephone system, you either had to pay a franchise fee to American Bell Telephone or invent your own equipment that didn't conflict with any Bell patents. Otherwise, you had to wait until 1893 when the initial Bell patent would become public property.

Telephony always fascinated Nathan more than telegraphy. While the telegraph was a mature technology, the telephone industry still had room for innovators and entrepreneurs. He could readily learn about the technical aspects from the periodicals. He found that he not only understood the descriptions and diagrams but could replicate the equipment and make it work. He also had time for experiments, especially in the winter when his gardendid not demand his attention. Here was a growth industry into which he could channel both his inventive energy and his desire to earn his own fortune.

The easiest way to circumvent the Bell patents was to build a non-electric, acoustic telephone system. Similar to the tin cans and string devices that kids use at play, these telephones were actually quite prevalent in the era. The US patent office issued dozens of patents for variations of the technology in the 1880s. One of those, no. 378,183, went to Nathan B. Stubblefield and Samuel C. Holcomb on February 21, 1888.

Nathan called it a mechanical or vibrating telephone. It consisted of a tin can mounted securely on a wooden backboard with a small round hole aligned with the center of the can's open back. A cloth diaphragm stretched across the open face of the can. In the center of this diaphragm was a button attached to a wire that led through the hole in the back of the device. The wire connected each phone to its mate. A coat of varnish on both sides gave the diaphragm structural integrity so that the installer could tighten the wire and stretch the surface, like tuning a drum. Finally, a small hammer hung from the device. To make a call, you hit the button with the hammer to signal the other phone, then stood close to the diaphragm and spoke.

Because wire was expensive, Nathan tested an early mechanical telephone connecting the instruments with waxed string. When he told Duncan Holt in 1885, "I've been able to talk without wires ... all of 200 yards ... and it'll work everywhere," Nathan was probably talking about the mechanical telephone, which technically was "wireless." Subsequently, writers picked up that anecdotal comment and interpreted it to mean that Nathan already was at work on the electrical wireless telephone systems, but it is unlikely that he had access to key components in 1885. Besides, Nathan stated that his wireless efforts began around 1890.

Soon he abandoned the string connections in favor of wire that was more resistant to weather damage. Wind created particular problems for this type of phone system. It stimulated vibrations in the wires. Because the system was acoustic, the vibrations were audible, sometimes loud and piercing. More expensive copper wire reduced but did not eliminate the noise. Moreover, the distance between telephones was limited to short lines, not over a mile, because the connecting wire had to be stretched taut. But within those limits, Stubblefield claimed that the system was "capable of transmitting a whisper" and voice or music would have "such clarity as to be heard 100 feet away from the phone." He offered a five-year warranty on both equipment and installation.

Despite its low-tech design, this was the first telephone system in Murray, and Stubblefield experienced enough prosperity from its novelty value to establish an office on the town square. He and his partner Samuel Holcomb sold the first pair to Calloway County Court Clerk George W. Craig in late 1886. Craig wrote:

> This is to certify that I have had in use some months one of the Stubblefield & Holcomb Telephones which is giving satisfaction. Therefore I do not hesitate to recommend it.

His affidavit is one of several that Stubblefield collected for promotional use. Early customers attested that it was the best telephone that they had ever heard, perhaps because it was the only one they had ever used. A declaration from druggist A.H. Wear stated:

> This is to certify that Messr's Stubblefield and Holcomb have put up for us one of their telephones (the distance being about 3/4 of a mile) which works splendidly. When there is no wind blowing a whisper can be heard perfectly distinct. This is to our knowledge the best vibrating telephone we have seen.

Murray druggists Martin and Dale were also satisfied customers:

> We take the pleasure in recommending the Stubblefield and Holcomb Telephone. We have two of them in our store, one leading from our store to office above with four angles; and the other a residence some 450 yards distant. Both work satisfactorily.

Samuel Holcomb's role in this enterprise remains obscure. No relation to a Holcomb who was in a similar business in Cincinnati at the same time, Samuel probably put up the money for the patent application and served as a field sales representative.

Stubblefield sold and installed the telephones himself around Murray and nearby West Tennessee. He also ventured farther south in 1887. One customer was John Gage, a merchant in Louisville, Mississippi. Nathan described him:

> This man had a hereditary trait of urbanity in him … a fat belly man, jolly and clever. When his

> phone was put in it carried such an interest with it up in his end of town that in order to hear the wonderful thing talk the people came in by the score and they came into his parlor with mud on their feet until they liked to have ruined a nice carpet.

On the same trip, he installed a system for the Illinois Central Railroad at McCool, Mississippi where the railroad agent wrote:

> Mr. Nathan Stubblefield of Stubblefield and Holcomb Murray Ky. has just completed me a telephone line which works admirably. Have tested the line by playing harp or whispering at one office which can be distinctively heard and understood at other office. Anyone needing a line can be accommodated by Mr. Stubblefield who guarantees satisfaction and has given it here.

By January 1888, Nathan and his vibrating telephone were well known in and around Murray. Calloway County Judge W.B. Keys, along with other county officials, signed a testimonial that stated:

> This is to certify that Nathan Stubblefield of this town has from time to time during the last 14 months put up quite a number of the Stubblefield Telephones in this town and county which are giving general satisfaction. ... We cheerfully recommend them to the public.

For three months in that spring, Stubblefield returned to Mississippi and set up shop in Vaiden “at the best hotel in town.” Numerous affidavits attest to the success of this trip.

Stubblefield had a territorial deed for any representative who wanted to establish a local franchise elsewhere using his system. One agent from Tennessee, G.G. Westerbrook, was successful at selling the “Vibrating Telephones” as far away as Oklahoma. J.T. Stubblefield had a franchise in the Pacific Northwest. The most suc-

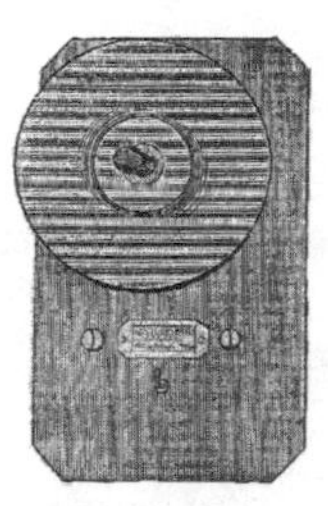

TERRITORIAL DEED.

Stubblefield's Patent Vibrating Telephone,

N. B. STUBBLEFIELD, Prop'r, Murray, Ky.

WHEREAS, I, Nathan B. Stubblefield, and Samuel C. Holcomb, of Murray, county of Calloway, State of Kentucky, did obtain letters patent of the United States for Vibrating Telephone, which letters patent are numbered 378,183, and bear date the 21st, day of February, in the year One Thousand Eight Hundred and Eighty-eight, and, whereas, I am the sole owner of said patent and of all rights under the same in the below recited territory; and, whereas, ...,

of ...., county of ...., State of ....

is desirous of acquiring an interest same: Now, therefore, to all whom it may concern, be it known that for and in consideration of the sum of $ .... to me in hand paid, the receipt of which is hereby acknowledged, I, the said Nathan B. Stubblefield, have sold, assigned, transferred, and set over, and by these presents do sell, assign, transfer, and set over to the said ...., all the right, title, and interest whatsoever which I have in and to the said invention, as secured to me by said letters patent for, and in the ...., and for, to or in no other place or places; the same to be held and enjoyed by the said ...., within and throughout the above specified territory, but not elsewhere, for his own use and behoof, and for the use and behoof of his legal representatives, to the full end of the term for which said letters patent are granted, as fully and entirely as the same would have been held and enjoyed by me had this assignment and sale not been made.

IN TESTIMONY WHEREOF, I have hereunto set my hand and affixed my seal, this .... day of ...., A: D., 189....

In presence of

....

....

MURRAY (KY.) LEDGER STEAM PRINT.

Pogue Library

Exclusive Franchise Agreement for the Stubblefield Vibrating Telephone, 1888.

cessful representative was Ira Prichard of Murray, who wrote colorful accounts for the *Murray Ledger* of his sales trips throughout Southern Illinois in early 1889.

Installation cost varied, depending on the length of the line and the number of poles involved. But money was not common in the rural United States of the late 19th century. One customer, a Post Office in Coffeeville, Mississippi, came up short on a seventeen-dollar installation. Stubblefield had to make special arrangements. "These people liked [sic] about four dollars paying for this but it was paid for by public subscription, Mr. Brown taking responsibility of collection. So I let him beg off a little." No accurate count of the total number of telephones sold exists, but the income was adequate for Stubblefield to earn a living from the enterprise for at least four years.

After Stubblefield received his 1888 patent, he devised an improved version called the Laryngaphone. With it, the caller could communicate without the phone box by stretching a string tightly around his throat or clinching it in his teeth while talking. It used a hearing tube attached to the side of the metal can for better audibility and included an electric bell for better signaling. Although this electrical circuit would increase installation and maintenance cost, the new network would put Stubblefield in better position to enter the electric telephone business in 1893, when both his five-year warranty and the initial Bell patents expired. In preparation, he acquired copies of the relevant patents, contacted equipment manufacturers for price lists, and got legal opinions on the validity of claims made by the Bell Company and its subsidiary Western Electric.

Providing telephone service to the Murray newspaper undoubtedly helped Stubblefield get favorable publicity. One article about his work read:

> He is an inventive genius, and his inclinations are to experiment with electric appliances, etc. ... Mr. Stubblefield has about reached perfection in the manufacture of telephones and has his lines introduced throughout a large scope of territory. He SAYS he intends building up a "telephone rep-

Wrather Museum

Stubblefield's Laryngaphone, 1890
Never patented, this was a variation on the Stubblefield Vibrating Telephone.

utation" over the name of Nathan Stubblefield.

The editorial support from the *Murray Ledger* was welcome because Nathan now had competition in the telephone business. In 1889, a group of Murray doctors and businessmen, many of them Stubblefield's subscribers, formed a company to bring the Bell telephone system to town. With the latest improvements in sound reproduction technology, it was now superior in every respect to Stubblefield's vibrating telephone. While Stubblefield was marketing a novelty, his competitors sold a versatile service that had better prospects for return on investment.

A severe technical limitation of acoustic telephones was that they were closed circuit systems — one phone connected to one other — without provision for central switching or networking. Customers needed a separate telephone and line for each location that they might want to call. Economically, this flaw limited Stubblefield's income to the initial sale of equipment and the low-margin business of maintaining existing installations. Because he could not establish local telephone exchanges, he had no opportunity to earn the substantial cash flow from the monthly billing for connections to local and, eventually, long distance networks.

Within a year Stubblefield's telephone business was finished. Understandably upset at the prospect of failure just as he was beginning to show results, he also knew that he had expertise that was valuable to his competitors. So he entered into a contract with his competitors to install Bell telephones, and a system including five miles of wire, a switchboard, and central office. But this arrangement proved unpleasant, and by late 1890, Nathan found himself back on the farm.

He made a few more attempts at wired telephony. The first was a novel telegraph device designed to work on telephone lines. It was unique in that users could dial up letters and numbers, rather than using Morse code, to communicate with each other. Similar devices, some with printers, experienced modest success in Europe but never caught on in the United States. Like the Laryngophone,

Wrather Museum

Stubblefield's Bell Telegraph, 1890
Designed for home use, this device utilized letters of the alphabet rather than Morse Code. He never applied for a patent on it.

Stubblefield's Bell Telegraph didn't get beyond the design stage. Then in 1894, he tried to bring the Viaduct electric telephone to the area in competition with the Bell system. For this he had grand plans, beginning with replacing the few mechanical phones that he still maintained in Murray with the new electric ones. Then he envisioned franchises in every state and territory for the Stubblefield Telephone, with prices ranging from $200 in Arizona and Montana to $5000 in New York. But his competitors had already sucked up both capital and customers. A charter from the city of Murray was the extent of this enterprise, and he finally sold that for $50 to pay off debts.

From his experience in the telephone business, Nathan learned two lessons that would stay with him for the rest of his life. He began to think on a grand scale of inventions that would have a nationwide market. And he discovered that even people he had known and trusted all his life would sell him short in a minute to make a profit. So when he had any grand ideas, he'd best keep the details to himself until they were ripe, like the melons in his garden, and ready to sell.

In the meantime, Stubblefield had encountered a technology that he felt would allow him to recoup his losses. With it he believed that he could establish a telephone system at substantially lower cost than any Bell franchise and provide service to more customers, even those widely dispersed throughout rural America. It was a wireless telephone system, and it would make Nathan Stubblefield the most famous person ever to come from Murray, Kentucky.

Wireless in 19th Century America

The task that Nathan Stubblefield faced — to create a practical wireless communication system — was not new. Similar challenges had confronted telecommunications inventors for half a century. They employed a variety of electrical technologies, including natural conduction, modulated light, induction, and finally electromagnetic waves. The first to come up with a solution, using the natural conduction method, was Samuel F.B. Morse, and he did it almost by accident.

It was an embarrassing distraction that Morse didn't need. In 1842, he had enough trouble trying to create interest in and raise capital for his great obsession, the telegraph. But Morse had to face the facts — his submarine cable demonstration in New York harbor had been a failure.

It had been seven years since he built his first working telegraph. So far he had been unable to convince either investors or politicians of the invention's potential value. Now he faced another barrier, a physical one. It was comparatively easy to build an overland telegraph. All you needed was a right of way for poles or a trench to run the wire. But what happened when you reached a body of water too wide to span with a single run between two poles? Morse reasoned that the best solution would be to insulate the wire and run it underwater.

He chose New York harbor to test his idea. There he could also attract attention and perhaps investment. He planned to transmit and receive messages between the Battery, at the south end of Manhattan Island, and Governor's Island, about 1 mile distant. The first challenge was to make a waterproof cable. Wire itself was a rare commodity in 1842, so he first had to find a metalsmith who could draw a strand of 2 miles continuous length that the project required. Then Morse had to wrap it by hand "with hempen threads well saturated with pitch, tar, and surrounded with India rubber." Then he had to carefully coil the brittle cable so as not to damage the insulation nor break the wire and load it into a rowboat.

Next, Morse and an assistant had to row across the channel while slowly unrolling the cable and letting it settle to the bottom with enough slack so that it would not be an obstacle in the shipping lane. As they went along, they inspected every inch for cracks in the insulation, patching the cable with raw rubber everywhere that the corrosive salt water might seep in. They worked all day and well into the night. Finally, the task complete, Morse rowed back to the Battery, and they tested transmission and reception. It worked.

The next morning, October 19, the *New York Herald* announced the demonstration:

> This important invention is to be exhibited in operation at Castle Garden between the hours of twelve and one o'clock today. One telegraph will be erected on Governor's Island, and the other at the Castle, and messages will be interchanged and orders transmitted during the day. Many have been incredulous as to the powers of this wonderful triumph of science and art. All such may now have an opportunity of fairly testing it. It is destined to work a complete revolution in the mode of transmitting intelligence throughout the civilized world.

With such a build-up, it's no wonder that a crowd of curious onlookers had assembled by the time Morse arrived at mid-morning. With Leonard Gale manning the instrument on the island, Morse commenced his demonstration. He sent a few characters and received a few back. Then, as the instrument was in the midst of punching dots and dashes into the paper tape, the line went dead. Unable to restore the circuit, Morse had to cancel the demonstration, much to the delight of the derisive and jeering crowd. What went wrong?

Peering out into the harbor, Morse saw the answer. Several ships were anchored between the Battery and Governor's Island. One of them had weighed anchor and hooked the cable in the process. Mistaking it for a rope, the sailors had cut it away. Morse had no immediate way to repair the physical or the public relations damage. Fortunately, since there was yet no telegraph to disseminate the news widely, the negative publicity was confined to New York.

Morse licked his wounds and pondered the problem for the next few months. He came up with a novel solution — a wireless telegraph. In Germany, Sömmerling and Steinhill had shown that water or earth could serve as conductors for an electrical circuit. Telegraphers soon replaced the second wire in the system with a ground connection. Morse reasoned that if a body of water could furnish the return circuit for his telegraph, it could also serve as the primary, thus eliminating the need for wires period.

In December 1842, the inventor devised an experiment across a canal in Washington DC, where he lived. He used two wires, one attached to a telegraph key and a battery and the other to a galvanometer to detect changes in the voltage. The ends of the wires were fastened to metal plates that were submerged on opposite banks of the canal. He tried the device on December 16, and two days later wrote to his brother Sidney:

> I believe I drew for you a method by which I thought I could pass rivers without any wires through the water. I tried the experiment across the canal here on Friday afternoon with perfect success; this also has added a fresh interest in my favor, and I begin to hope that I am on the eve of realizing something in the shape of compensation for my time and means extended in bringing my invention to its present state.

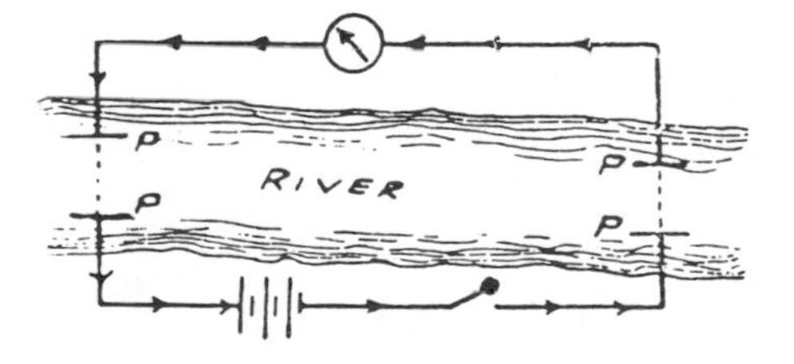

Morse 1842
Wireless
Telegraph System

As Morse had expected, the water completed the circuit. Although this span was only 80 yards, the next year his assistants successfully transmitted messages across the Susquehanna River at Havre de Grace, Maryland, a distance of 1 mile. They learned that larger metal plates, more widely spaced, increased both transmission distance and signal quality.

Morse finally received a government grant in 1843 to build a wired telegraph line from Washington to Baltimore. From that point, commercial development of this system occupied his time. He never applied for a patent on the natural conduction wireless telegraph, but its design tantalized electricians for many years. James Bowman Lindsay, a Scottish inventor, tested a Morse-type wireless telegraph in 1854 and even proposed a grand but impractical installation for Trans-Atlantic service. The completion of the undersea cable link from England to Canada in 1866, however, ended what little interest Lindsay had stimulated.

Mahlon Loomis, a dentist from Washington DC, believed that he was using natural conduction when he created a wireless telegraph in the mid-1860s. His system consisted of two lightweight metal grids hoisted on kites from hilltops. Metal wire connected the grids to the ground. Since the moist upper atmosphere was a superior electrical conductor and the electricity there had an opposite charge from that of the earth, Loomis figured that his system merely completed a circuit. All he needed was one device to interrupt the current, like a telegraph key, and another to detect the interruptions, like a galvanometer, attached to the kite wires. In 1866, he tested this apparatus successfully at a distance of 14 miles, and received a US patent on it six years later. He later made unsubstantiated claims that he used the same system as a wireless telephone.

What Loomis actually accomplished remains an enigma. Some historians have suggested that, by flying the kites at the same altitude, he unwittingly created an antenna system that utilized tuned resonance, a key component of radio. Others think that he was probably detecting atmospheric phenomena rather than telegraph signals. With no patent model or illustrations, few detailed notes of his experiments, vague affidavits, and one box of random pieces of equipment in storage at the Smithsonian, Loomis left very little evidence behind.

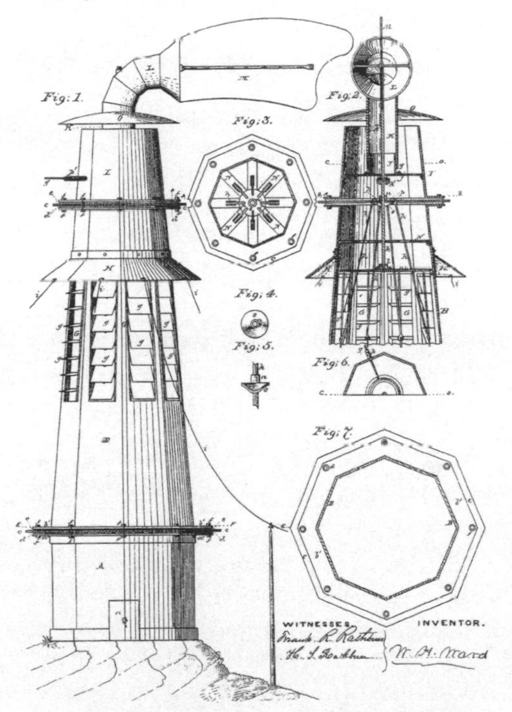

Wireless Telegraphy by Convection

William Henry Ward, 1872

An amusing sidelight to the Loomis experiments was a US patent granted to William Henry Ward of Auburn, New York. Predating Loomis' by three months, this was technically the first US patent for a wireless telegraph device. Ward was a shameless huckster, author of religious tracts, and creator of inventions ranging from naval signaling systems, to a self- regulating bombshell fuse, a bullet compressing machine, and hair pomade. He met Loomis when the dentist consigned a set of patented false teeth to Ward for display at a trade show in England. Ward apparently learned of the Loomis wireless telegraph and decided to try his hand at a similar invention, with absurd results. The patent "Improvement in Collecting Electricity for Telegraphing, &c." shows a vented tapered tower with a horizontal vane at the top that rotates in the wind. Constructed of two dissimilar metals, the vane produces a low-voltage electric signal in moist air. A combination of wind and hot air rising through the core of the tower would propel the signal to the next tower — wireless by convection.

Just as Morse and other early telegraphers faced the challenge of wireless communication so too did telephone pioneers like Alexander Graham Bell. Inspired by geographers who had made precise measurements by sending electric currents through the earth and water, Bell created a refinement of the Morse wireless for experimental communication between two boats. He tested this telephone system on the Potomac River near Washington in 1879, at distances up to a mile and a quarter, with only marginal results. Probably the poor quality of the telephone transmitters was at fault.

Bell dropped this line of investigation for two reasons. He learned that John Trowbridge, a physics professor at Harvard, was testing a similar system. But more importantly, Bell had a better idea. He wanted to invent an optical wireless telephone using the naturally photoelectric element selenium in the receiver to turn light into an electric telephone signal. Working with Charles Sumner Tainter at his lab in Washington, Bell spent two years perfecting the invention that he named the Photophone, awarded first US patent for a wireless telephone in December 1880.

The major problem that Bell and Tainter encountered was transmitter design. They had to devise a way that a voice or other sound could modulate a beam of reflected light. Their solution was a silver coating on the back of the vibrating diaphragm in an acoustic telephone transmitter . With precise alignment, a beam of sunlight passed through a lens that focussed it onto the back of the diaphragm, that in turn directed it to the parabolic reflector whose center point was a piece of selenium connected to by wire to a battery and a telephone receiver. Whenever someone spoke into the transmitter, the diaphragm vibrated and made the light flicker slightly. These fluctuations in light intensity caused proportional changes in electric current flowing through the selenium. The telephone receiver turned the current modulations back into sound.

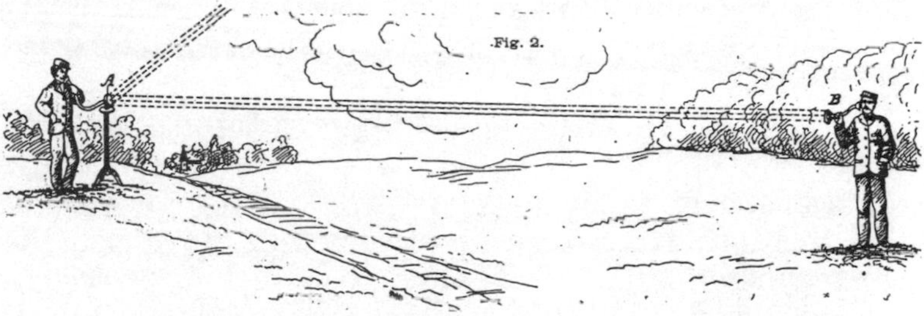

The Photophone, Alexander Graham Bell, 1880
First wireless telephone patented in the US

A later version of the invention used radiant energy rather than light as the transmission medium. A French electrician named this the Radiophone, the first use of the term *radio* for a wireless telephone. Also, Western Electric engineer Lloyd Espenschied pointed out that with its claim to modulate radiant energy by sound, Bell's invention was the basis for AM radio. Although the Photophone was well publicized and the subject of experiments and improvements for more than twenty years in Europe, it never progressed beyond the novelty stage. The mechanism was too complicated and delicate, and the Photophone never achieved substantial transmission distance. Shortly after American Bell Telephone decided not to pursue its development, Bell ceased active involvement with the company and gave the models of his invention to the Smithsonian.

An electric current passing through a coil of wire presents three possibilities for wireless communication. First the current generates static electricity that may, under certain conditions, discharge as a spark. Second the current generates a magnetic field around and perpendicular to the coil. This field will in turn stimulate a continuous current in a second separate coil nearby. Variations in the current through the first coil will create proportional variations in the current through the second. This is *electromagnetic induction.* Its discovery is jointly attributed to Michael Faraday in England and Joseph Henry in America in the 1830s. In 1842, Henry conducted a series of experiments at Princeton called "Induction at a Distance, " during one of which he detected wireless electricity from a coil two hundred yards away.

In the third effect, the electric current through the first coil touches off electromagnetic waves that radiate through the atmosphere. When one end of the coil is grounded, and the other terminates in an elevated metal plate or grid, the wireless communication is more efficient. This is *electromagnetic radiation.* Although the waves propagate to distances that far exceed the induction field, they require a more sophisticated detector at the receiver. Scottish mathematician James Clerk Maxwell, in his 1866 treatise on the nature of light, predicted the existence of electromagnetic waves at various frequencies and wavelengths. Thomas Edison and Elihu Thompson in the United States, and David Hughes in Britain, encountered unexplained wireless electric phenomena in the 1870s that were probably radiated waves. Edison called it an "etheric force." In 1888, German scientist Heinrich Hertz finally proved the existence of electromagnetic waves by building equipment to transmit and receive them in his lab.

Electricians on both sides of the Atlantic were involved in experiments with wireless induction communications. American inventors seemed to make more progress in the 1880s. Thomas Edison, Lucius Phelps, and Granville Woods all received patents for wireless telegraphs, and Amos Dolbear secured one for a wireless telephone based on induction.

Dolbear, a professor at Tufts College in Massachusetts, pursued the invention of the electric telephone simultaneously with Bell. He also invented an electrostatic telephone receiver that had superior sensitivity but was too delicate and expensive for regular use. Western Union paid Dolbear $10,000 for his telephone patents and used them in an unsuccessful patent infringement suit against the Bell Company. By 1881, the professor was comfortably ensconced in his laboratories, happy to spend the rest of his life as a teacher, writer, and scholar. He discovered wireless telephony quite by accident. Dolbear described the event:

> While at work at the single terminal receiver..., the cord became detached from the line while I was unaware of it, and I still heard the speech from the transmitter plainly. Upon noticing this I began backing away from the end of the wire from the transmitter, letting the single cord hang free in the air. I could hear the talking in the most remote part of the room...

Dolbear filed a patent application on his "Mode of Electric Communication" in March 1882 and received the patent four years later, the second for a wireless telephone. He demonstrated the device widely, including presentations at the Society of Telegraph Engineers and Electricians in London, 1882, and the American Association for the Advancement of Science in Montreal, 1883. At the London venue, Dolbear hired a trumpet player to send live music by wireless telephone to the listeners in the next room.

Soon after Dolbear's wireless telephone patent was

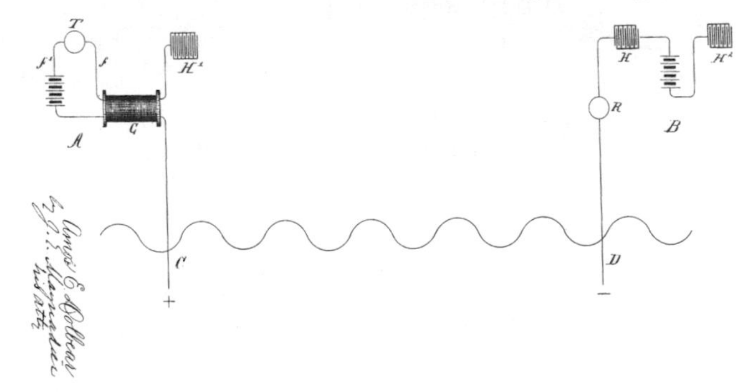

Induction Wireless Telephone, Amos Dolbear, demonstrated publicly in the US, Canada, and Europe in 1882 and 1883, patented in US, 1886

issued in 1886, *Scientific American* published a detailed article, complete with illustrations and technical explanations. Electricians later remarked that the Dolbear system, which utilized elevated, grounded condensers as antennae, actually transmitted electromagnetic waves. It only lacked a detector at the receiver to be a complete radio system. Dolbear never attracted capital to develop his invention commercially.

Creating a telegraph system to communicate with moving trains occupied several inventors in the 1880s. In 1880, there were more than 8,000 train wrecks in the United States. Having instant communication with the trains might avert much loss of life and property. Granville Woods of Cincinnati was apparently the first to devise a wireless induction telegraph for this purpose in early 1881, although he didn't apply for a patent until 1885. William Wiley Smith of Indianapolis actually received the first patent for such a device in 1881. Smith attempted to communicate with existing telegraph lines running parallel to the train tracks, but its design was inferior.

Lucius Phelps of New York probably advanced the technology further than his competitors, and certainly got the most publicity. Phelps applied for patents in late 1884, and installed a working system on the New York - New Haven Railroad in February 1885. By then Thomas Edison and Ezra Gilliland had acquired the Smith patent, revised it, and submitted a competing application in May 1885. Calling his invention the "grasshopper telegraph," Edison confidently predicted that it would be an indispensable communication tool for businessmen traveling by rail.

Both *Scientific American* and *Electrical World* covered the Phelps and Edison systems with numerous illustrated articles in 1885 and 1886. Apparently, Granville Woods read one of these, saw that someone was about to profit from his idea, and rushed to a patent attorney with his sketches, models, and affidavits from witnesses in 1881. Once Woods filed his application for a wireless telegraph, also in May 1885, the Patent Office declared an interference and scheduled two sets of hearings, one to settle the Phelps - Edison dispute and another to settle the Phelps -

coil and hide it among branches and leaves? So long as the listener stayed inside or near the transmission coil and the two coils remained roughly parallel, Nathan would be able to transmit freely to the portable set as he walked around. This design is similar to Dolbear's "accident" in his lab ten years earlier. The hidden wire that connected Nathan's transmitter to the coil could stretch as far as he wished, but the actual distance between the two coils would actually be quite short.

Mason mentions "a keg with a handle on it." A round vessel is a very convenient place to put a coil of wire, especially one with many turns. The more turns in the coils, the stronger the inductive effect. Since Mason doesn't mention talking back, this may have been an experimental receiver.

We also get a glimpse of the man behind the technology in these tales. This is Nathan with his Derby hat showing off his produce, not a reclusive truck farmer. It's also Nathan the genius matching wits with a precocious kid and a learned doctor and coming out on top, perhaps even capable of a harmless practical joke.

Over the next few years, Nathan's demonstrations became more numerous and more public. One account mentions wireless transmission from one floor of a building to another, and several people affirmed that by 1898, they had witnessed and used Nathan's wireless telephones at his home, on the courthouse square, and elsewhere.

At some point, Stubblefield moved away from induction technology. Unfortunately, he learned what others had. Induction wireless telephones required enormous coils and considerable voltage to achieve even modest transmission distances. They were unwieldy and impractical as alternatives to a wired telephone system. After his telephone business failed, Nathan still had a small stockpile of equipment but was short of money to buy consumables like wire and batteries. So he began to build his own batteries by wrapping copper wire around a soft iron core, then burying the device in the ground where moisture served as the electrolyte for the reaction between the two metals.

....Stubblefield's....

Electrical Battery.

[No. 600,457, Patented March 8, 1898.]

Following is a brief description of New Electrical Aparatus recently patented by us in the United States, with various practical applications enumerated.

Fig: 1.

PATENTS PENDING IN ENGLAND AND CANADA.

Figures 1 and 2 show in outline the general appearance of our cell, which in construction is simply a solenoid of copper and iron wire, the seperate elements wound close and compactly about a soft iron core. As will be seen, the negative and positive elements (coppei and iron) lay close together, but are insulated from the core, and throughout the entire length.

Non acidulated water or merely moisture constitutes the electrolyte, when used as other types of battery, though when used as an EARTH CELL in addition to the moisture of the earth, it is subjected to some electrical action of the earth's charge not very well understood, but presently to be further described.

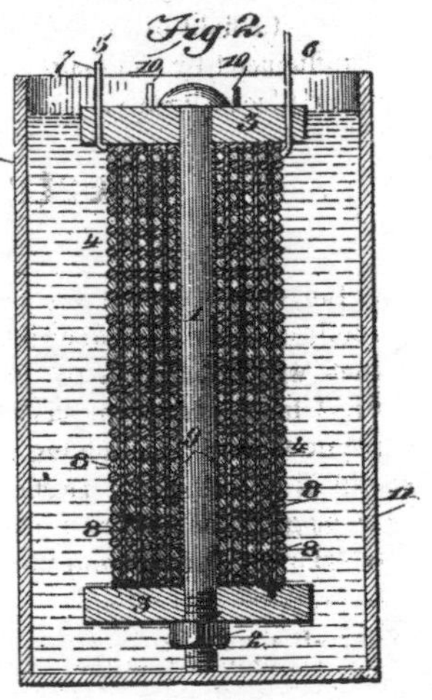

When coil or cell is placed in jar as seen in figure 2 a practically constant electro motive force of something less than one volt is the output, the cell being practically free from polarization effect common with most types of battery, (cell as shown in figure 2 may be connected up in series to obtain any required voltage.)

At first thought it would seem that this cell was impractical, cost of installation considered but costs of maintainance and renewals should be the points of economy to be noticed, no attention

Pogue Library

Promotional Brochure for Stubblefield's Electrical Battery or Earth Cell, 1898

Pogue Library

Electric Motor that Stubblefield operated non-stop for 66 days with one Earth Cell battery. There is no explanation for the telephone receiver.

Eventually, Nathan invented what *Electrical World* called "a primary battery of novel design," for which he received US Patent 600,457 in 1898. In a promotional brochure, he claimed that the battery was also a self-generating electromagnet that could be turned into an induction coil with the addition of a secondary wire wound around and insulated from the primary coil. "As an induction coil the economic applications should be many," he said, "as this also can be placed in the earth with proper protection provided for the secondary wire, no attention afterwards being necessary." To prove his point about perpetual motion, Nathan built a small electrical motor and operated it for more than two months non-stop with one of his earth cell batteries. The brochure also proclaimed the device's usefulness to power relay bells, electric lighting, overland telegraph stations, and ordinary telephones. It was "of especial value in electro-theraputics [sic]."

In order to raise capital for the patent application, Nathan sold a half interest in the earth cell battery to William G. Love, a photographer from Murray. Love also took many promotional photographs through the years for Stubblefield, so the patent interest was perhaps a payment for services. Although Nathan never sold this battery widely, its invention was an important step in several respects.

In a section of the battery brochure called "Telephoning Through the Ground," Nathan wrote:

> With coil … placed in the ground, and properly connected the proper transmission of intelligence of sufficient volumn [sic] of sound for commercial purposes is had, using only bear [sic] wire through ground or water as line of transmission. Tests as to distance, 3,500 feet only, yet have been made, but sufficient evidence on investigation will leave no doubt as to the accomplishment any distance, as the inventor here again reasons that the electrical energy of the earth is the prime cause or factor. (In this experiment the microphone transmitter is used with a slight modification of the telephone induction coil.) This we con-

> sider a special point in favor of our earth cell, and of special commercial value as thus the dangers of lightning or the inroads of malicious persons are obviated. For Exchange work or any arrangement that this presents a practical front in point of economy over the cable or conduit systems well known, as well as in emergency cases quick equipment for subteranean [sic] or submarine telephony may be installed. Telegraphic communication with the Morse instrument is practical through the ground, for distance of a few thousand feet through the medium of this earth cell as described, but as to the limit of such a system of telegraphy we are not yet prepared to state.

Nathan shared the common misconception that he was drawing electricity from nature rather than generating it with a chemical reaction. His approach to natural conduction wireless communication was consistent with what Morse had used in 1842 and with Bell's experiments 20 years earlier. With the earth as his transmission medium, Nathan found that he could easily double the maximum transmission distance of his induction wireless. This was his direction for the next four or five years.

Although he probably preferred to work alone, Stubblefield realized that it was impossible to conduct his experiments without someone on the other end of the telephone. By 1900, Nathan's oldest son Bernard was 12 and had become his able and reliable assistant. There is some evidence of numerous public demonstrations in and around Murray in 1900 and 1901, with most anecdotes placing these events on the courthouse square, perhaps on market day. Witnesses generally described the apparatus as having telephones attached to boxes whose contents were not divulged, wires connecting the boxes to a pair of metal rods that were sunk in the ground. Bernard usually operated the transmitter, using a set routine of counting to ten, whispering, whistling, and blowing on a harmonica. Having discovered that receivers worked almost anywhere he could drive the ground rods, Nathan often set up multiple listening posts for simultaneous reception.

Although no newspaper accounts of these events survived, Nathan apparently developed quite a flair for publicity. On January 1, 1902, the *Paducah Sun-Democrat* published the following:

> Nathan Stubblefield, the Calloway County man who claimed to have discovered perpetual motion a few years ago, is now being exploited in some of the newspapers as having solved the problem of wireless telegraphy, and that by his plans a transmitting apparatus of no gigantic magnitude may be placed at some central part of the United States, with proper earth connections, and the signal service flashed to all parts of the country, and says that at a cost of a few dollars each home in the land may be equipped with a receiving apparatus, and thus with the earth connection receive the weather forecasts and other news of the day, transmitted from this central point with all the ease of telegraphic communication. Well, Nathan is a genius, and ought to do most anything if he can utilize the forces of nature as successfully as he can work the newspapers.

On that same New Years Day, Nathan and Bernard put on their grandest wireless show to date in Murray, demonstrating not only the practical usefulness of the technology but its broadcasting capability as well. With the transmitter on the courthouse square and Bernard going through his prepared audio routine, listeners at five receivers around town heard the results of what was quite possibly the first wireless broadcast. In all, at least 1000 people attended the event. The *Louisville Courier-Journal* published, and subsequently syndicated, an account the next day.

> Since the public test of wireless telephony by inventor Nathan B. Stubblefield, of this country, and his demonstration that he could transmit the sound of the human voice over distances of

Pogue Library

Nathan and Bernard Stubblefield,
with the wireless telephone demonstrated on the
Courthouse Square, Murray, January 1, 1902

> ground without wires, the eyes of the academic world have been turned upon Murray.

Nathan also expounded to the reporter on his invention's usefulness in disseminating weather and other emergency information, as a means to communicate between ships at sea, and a simple inexpensive system to establish telephone service in rural America. Although he erroneously believed that his invention tapped a reservoir of electricity the same way a well digger found water, his concept of the physics of electricity was not that far removed from mainstream science. A few days later, the *Louisville Courier-Journal* asked the chairman of the physics department at Kentucky State College to explain how Stubblefield's system worked. Without ever laying eyes on the equipment, he said:

> Instead of using a wire to transmit sound, the ether may be used, and the electrical energy may be transmitted in the form of ether waves. The ether is the great vehicle for the transmission of energy. This medium fills all space, interplanetary and intermolecular.... This same ether is electricity and ... all electrical phenomena are due to some disturbance of the ether. The ether is easily thrown into vibration resulting in ether waves. There is an immense variety of these waves ranging in length from a few millionths of an inch to hundreds of miles. Some of these waves affect the eye and are called light waves; some transmit heat energy. They are all electromagnetic waves and all travel with immense velocity.

In January 1902, wireless was still very new. Marconi, Lodge, Popov, Slaby, and others had only scratched the surface of the possibilities for Hertzian wave telegraphy. Their equipment lacked amplification that would make voice transmission possible; the spark transmitters emitted too much audible static; and the tuning mechanisms were rudimentary at best. Less than a month before Nathan's Murray demonstration, Marconi had finally succeeded in sending one letter of Morse Code across the Atlantic. Even trained scientists and electricians didn't fully understand the differences between the various modes of wireless transmission, so it's understandable that journalists and the general public would be ignorant of the arcane details as well. To them, Stubblefield had created something special and unique, and that was big news.

CATCHING WIRELESS MESSAGES A MILE BACK IN THE WOOD ON THE STUBBLEFIELD FARM.

Pogue Library

INVENTOR STUBBLEFIELD RECEIVING MESSAGES AT THE STATION 500 YARDS FROM HIS HOUSE.

Pogue Library

Nathan Stubblefield at work, 1902
Photographs from the *St. Louis Post Dispatch* feature

The Wireless Telephone Company of America

Kentucky Farmer Invents Wireless Telephone

Nathan Stubblefield Raises Vegetables for Market in Order That He May Live, But Has for Ten Years Devoted All of His Spare Time to Electrical Experiments, Until Now He Has Perfected a Wireless Telephone System Over Which Messages Are Distinctly Heard at a Mile.

St. Louis Post-Dispatch, January 12, 1902

The headline for the full-page feature in the Sunday magazine proclaimed all the excitement that the reporter felt. Picking up a lead from the Kentucky press, the editor of the St. Louis daily assigned an unnamed reporter to the Stubblefield story. On January 10, he arrived in Murray and received an eager welcome at the farm, the first newspaper correspondent to be invited for a private demonstration and tour. Once again, Nathan the salesman in the Derby hat appeared.

For the better part of the day, reporter tested the wireless telephone system, asking questions and taking notes along the way. For his part, Nathan realized that he was on the verge of national notoriety and was surprisingly candid in his responses. The only facts that he failed to divulge were the details of the invention itself, which was yet unpatented. Nathan claimed that these descriptions were in a document filed with the Calloway County court clerk and witnessed by several pillars of the community.

The reporter depicted Nathan and Bernard, their working conditions, and the experience itself with precision:

> He led the way to a tiny workshop built onto the porch of his house. It was just wide enough to hold the transmitter, which stood before the window, and a chair. One end of the room was given up to shelves laden with technical books on electricity.

The transmitting apparatus is concealed in the box before described. Two wires of the thickness of a lead pencil coil from its corners and disappear through the wall of the room, and enter the ground outside. On top of the box is an ordinary telephone transmitter and a telephone switch. This is the machine through which the voice of the sender is passed into the ground to be transmitted by the earth's electrical waves to the ear of the person who has an instrument capable of receiving and reproducing it.

The son of Mr. Stubblefield was left at the house to send the messages.

We went into the cornfield back of the house. Five hundred yards away we came to the experimental station the inventor has used for several months in working out his wireless telephony theory. It is a dry goods box fastened to the top of a stump. A roof to shed the rain has been placed on top of it; one side is hinged for a door, and wires connected with the ground on both sides run into it and are attached to a pair of telephone receivers. The box was built as a shelter from the weather and as a protection to the receivers. I took a seat in the box and Mr. Stubblefield shouted a "hello" to the house. This was a signal to his son to begin sending messages.

I placed the receiver to my ears and listened. Presently there came with extraordinary distinctness several spasmodic buzzings, then a voice which said: "Hello! Can you hear me? Now I will count ten. One-two-three-four-five-six-seven-eight-nine-ten. Did you hear that? Now I will whisper."

I heard as clearly as if the speakers were only across a 12-foot room the ten numerals whispered.

"Now I will whistle," said the voice. For a minute or more, the tuneless whistle of a boy was conveyed to the listener's ears. "I am going to play the mouth organ now, " said the voice. Immediately came the strains of a harmonica played without melody, but the notes were clear and unmistakable. "I will now repeat the program," said the voice, and it did.

Meanwhile, Mr. Stubblefield paced back and forth some distance from the station, calling occasionally, "Is he talking to you?" to which I nodded reply.

An examination of the station showed that the wires leading from the receivers terminated at steel rods, each of which was tapped with a hollow nickel-plated ball of iron, below which was an inverted metal cup. The wire enters the ball at the top and is attached to the rod. The rod is thrust into the ground two-thirds of its length. Another test was made after the rods had been drawn form the ground and thrust into it again at a spot chosen haphazard by the correspondent. Again, the "hello" signal was made by Stubblefield, and after a few minutes' wait came the mysterious "Hello! Can you hear me?" and a repetition of the program of counted numerals, whispers, whistling, and harmonica playing.

"Now," said Mr. Stubblefield who carried under his arm duplicates of the ball-tipped steel rods, "I wish you would lead the way. Go where you will, sink the rods into the ground and listen for a telephone message." Away we went down a wagon track, though the wide cornfield. A gate was opened into a lane between the hedge that bordered the field and a dense oak woods. We pursued the lane for about 500 yards and struck into the woods. I led the way. Into the heart of the woods we walked for nearly a mile.

In a ravine I stopped. "How far are we from the now?" I asked. "About a mile," Stubblefield

> answered. "Place the rods where you will and listen for a telephone message."
>
> I took the four rods from Stubblefield. Each pair of rods was joined by an ordinary insulated wire about 30 feet long, in the center of which was a small round telephone receiver.
>
> Two by two the rods were sunk in the ground, about half their length, and the wires between them hanging loosely, and with plenty of play. I placed a receiver at each ear and waited. In a few moments came the signaling "buzz," and the voice of Stubblefield's son saying "Hello! Can you hear me? Now I will count ten, etc." He went through the program heard at the station near the cornfield. The voice was quite as clear and distinct as it was 500 yards from the transmitting station.
>
> Stubblefield leaned against a tree, saying nothing, his arms folded, but with a look of triumph on his face. The deep silence of the woods made the mysterious voice with its message from a mile or more away, received by no visible means, seem eerie. Perhaps a look on my face prompted Stubblefield to say: "It makes me feel queer myself when I hear that voice come out of the ground, as often as I have heard it in our experiments."

Nathan encouraged the reporter to try different receiver locations. The test continued about an hour with the same productive results at each spot. Finally the pair returned to the house. At that point, the narrative became more subjective, lyrical and idyllic:

> On the way back to the house through the wood and field, Stubblefield told of his discovery. He bears the stamp of genius. He is a recluse. He has the thoughtful, absent air. He is eccentric. His neighbors shun him, while they respect him. None ever intrudes upon his privacy — they

know well that such an intrusion means a rebuff long to be remembered. The little town of Murray is full of stories of his eccentricities, which, possibly are the growth of not understanding a man whose ideal were far beyond those of the neighborhood. Where once they laughed at him and called him a crank, now they look up to him with awe. He has done something.

Stubblefield is 40, slender and of the well-to-do farmer type, but far above it mentally. For a livelihood he grows fruit — and the best in his section. His melons are said to be dreams of deliciousness. He protects his patch with electric wires, which announce to him the presence of intruders. Like other Kentuckians he knows how to use a shotgun. His melon patch and his orchard are, therefore, not often molested.

He comes from a family distinguished in his locality. His father was a lawyer, much respected in that part of Kentucky, and passing rich. His brothers are merchants, "well off," as the saying is, and leaders in the community. But Nathan Stubblefield is a man aloof. He cares only for his home, his family and — electricity. He educates his children in person and, after seeing that his family is well provided for, spends the remainder of his substance in electrical experiments.

His son, Bernard Stubblefield, 14 years of age, has for four years been his father's sole assistant. He is a remarkable boy. His father has been his only educator, and the lad is now an expert electrician and reads abstruse works on electricity and technical electrical journals with the same zest that other boys read stories of travel and adventure. His father says of the boy that he would be able to carry out and finish his system of wireless telegraphy should the father die, so closely has he been allied with every step in its discovery and development.

The article concluded with a lengthy statement from

Nathan himself, probably a press release, in which he talks about how his invention came to be and what its potential uses are. The entire text is attached as Appendix B. Aware of Marconi's accomplishments, he gave no evidence that he understood the Italian inventor's wireless system or the differences in that technology from his own design. Nathan implied that the distinction lay not in the technical details but in the choice of medium. "I have solved the problem of telephoning without wires through the earth as Signor Marconi has of sending signals through space. But I can also telephone without wires through space as well as through the earth, because my medium is everywhere."

Nathan claimed no more than a mile in transmission distance, but also mentioned a larger, more powerful apparatus in progress that would permit much wider coverage, claims similar to Marconi's upon his arrival in England in 1896. He again proposed using both air and water as conductive media, and spoke of wireless communication with ships, trains and other moving vehicles. He mentioned a planned trip to Washington DC to patent his invention before seeking capital to develop it commercially. Nathan seemed particularly fascinated with the broadcasting capacity of his wireless telephone:

> … it is capable of sending simultaneous messages from a central distributing station over a very wide territory. For instance, anyone having a receiving instrument, which would consist merely of a telephone receiver and a few feet of wire, and a signaling gong, could, upon being signaled by a transmitting station in Washington, or nearer, if advisable, be informed of weather news. My apparatus is capable of sending out a gong signal, as well as voice messages. Eventually, it will be used for the general transmission of news of every description.

But Nathan also acknowledged that this same feature made his invention less useful as a telephone, a flaw that needed attention:

> I have as yet devised no method whereby it can be used with privacy. Wherever there is a receiving station the signal and the message may be heard simultaneously. Eventually I, or someone else, will discover a method of tuning the transmitting and receiving instruments so that each will answer only to its mate.

For the next two months, Nathan prepared for the trip to Washington and his demonstrations there. Meanwhile, after the publicity in St. Louis and subsequently by wire service beyond the Midwest, news of his wireless telephone began to stimulate interest among prospective financial backers. He showed his local backers letters offering him substantial sums for the rights to his invention. But one promoter, Gerald Fennell of New York was more enterprising than the rest. According to Bernard, Fennell actually traveled to Murray with his wife, gained Nathan's confidence, and expounded on his plans to form a corporation and solicit capital investment based on the value of Nathan's wireless telephone. At the time, Stubblefield was noncommittal. He decided to make the Washington trip to see what other offers he might attract.

By mid-March, Nathan was ready to travel. He prefaced his journey with a series of press releases dispatched to major newspapers along the East Coast. The *New York Times* ran a front-page item on March 17, describing the earlier event on the Murray courthouse square:

> At his Kentucky home he placed the apparatus in the courthouse yard and receivers in the offices of the County Attorney, County Judge and Sheriff, and in two or three dry goods houses, all separated and not connected either with his central station or with each other by wires, and he says he talked to all points at one time.

Three days later, Nathan conducted his Washington public demonstrations.

For the event, he rented a small steamboat, the Bartholdi, from which he wanted to show how easy ship to shore communications were with his system. He chose a spot on the Potomac River across from Georgetown University, and set up a central station in a boarding house nearby. According to the *Washington Times* of March 21, the demonstration was an unprecedented achievement:

> A practical successful test of wireless telephony, made yesterday on the steamer Bartholdi, off the Virginia bank of the Potomac, may result in revolutionizing the present methods employed by local and long distance telephone companies. The tests on land demonstrated to a far greater degree the value of Mr. Stubblefield's invention. Communication was had at varying distances, and in every instance the tests were eminently successful. Mr. Stubblefield furnished convincing proof that he has achieved, in a basic form, something greater even than Marconi.

Pogue Library

The Steamer Bartholdi on the Potomac River, Washington DC, March 20, 1902
Stubblefield, third from left, rentedthis boat for a ship-to-shore wireless telephone test.

Pogue Library

NathanStubblefield receives a wireless telephone call, Washington DC, March 20, 1902. on the banks of the Potomac River.

The *New York Times* coverage, which was picked up in Kentucky by the *Louisville Courier-Journal*, was less flattering:

> A practical test of wireless telephony over a distance of a third of a mile was conducted on the Potomac River, just above this city [Washington], with partial success. A wire was grounded on the shore, and wires from a small boat stationed a third of a mile across the river were dropped from the stern. Those who participated in the test announced that they recognized the indistinct sound of a harmonica, and also heard human voices counting at the other end. Similar exper-

> iments overland were conducted with more satisfactory results. The tests were made under the direction of Nathan Stubblefield of Lexington [sic], Kentucky.

Waldon Fawcett, writing in the May 24th *Scientific American*, noted that this was almost the precise location where Alexander Graham Bell had tried a comparable experiment, with only marginal results, 24 years earlier. There's no evidence that Nathan knew of the coincidence. That article, the text of which is included as Appendix C, also mentioned the similarities between Nathan's invention and those of Sir William Preece in Great Britain and Prof. A. Frederick Collins in Philadelphia. Fawcett went on to describe the event in more detail:

> The most interesting tests of the Stubblefield system have been made on the Potomac River near Washington. During the land tests complete sentences, figures, and music were heard at a distance of several hundred yards, and conversation was as distinct as by ordinary wire telephone. Persons each carrying a receiver and transmitter with two steel rods, walking about at some distance from the stationary station were able to instantly open communication by thrusting the rods into the ground at any point. An even more remarkable test resulted in the maintenance of communication between a station on the shore and a steamer located anchored several hundred feet from the shore. Communication from the steamer to the shore was opened by dropping the wires from the apparatus on board the vessel into the water at the stern of the boat. The sounds of a harmonica played on shore were distinctly heard in the three receivers attached to the apparatus on the steamer, and singing, the sound of the human voice counting numerals, and ordinary conversation were audible. In the first tests it was found that conversation was not always distinct, but this defect was remedied by the introduction of more powerful batteries. A very interesting fea-

Pogue Library

Nathan Stubblefield, just to left of center, receiving wireless telephone call with witnesses, Washington DC, March 20, 1902. In the foreground, center, is one of the steel rods sunk in the earth.

> ture brought out during the tests mentioned was found in the capability of this form of apparatus to send simultaneous messages from a central distributing station over a very wide territory.

Eighteen persons posed with Nathan on the banks of the Potomac, but his caption, added more than nine years later, doesn't identify any of these witnesses by name. It only mentions that they "are people of Washington, New York, Chicago, Charleston, and Murray." Several are probably journalists. Among the others may have been potential investors and principals of the Gordon Telephone Company of South Carolina. Nathan had apparently agreed to sell that firm a system to establish service with offshore islands, but no records of this transaction or of the company itself survive. Bernard does not appear in any photos. The travel budget was too slim for him to accompany his father.

Nathan returned to Murray with his equipment, the photographs, press clippings, and a feeling of triumph tempered by reality. The Washington trip had been expensive, yet he still had not filed a patent application to protect his invention. The problem now was capital. He invited a group of local businessmen and their families to his home for a private demonstration and display of the artifacts of his recent trip. He carefully staged two photographs in the lawn beside his house, with the early spring daffodils and iris in the foreground.

The first shows the entire group of people. In front of them are Nathan's two transmitter/receivers from the Washington, a tray of common telephone batteries, a few Stubblefield earth cell batteries, and other pieces of electrical equipment. Two large coils of wire flank the array. They are stacked high with issues of technical periodicals, including *Electrical World*. To the extreme left of the photo is a black man, identified as Sam Stubblefield, holding two baskets of apples. George Gatlin who attended this event as a child later related the story of the apples and the odd small globes hanging from the bare tree limbs in the background:

A Celebration at the Stubblefield Farm, April, 1902, after the trip to Washington DC

Pogue Library

> Those who were present had written invitations. During the demonstration I was at the broadcasting station with Mr. Stubblefield's son and Sam, a Negro banjo player who was well known in Murray at the time. … A picture was taken of the group with apple trees in the background. The apple crop, of which Mr. Stubblefield was very proud, had been harvested, but for this special occasion apples were obtained and tied upon the trees so that they would show up in the picture.

Nathan's cross-promotional strategy is more apparent in the second photograph, a family portrait that frames the scene more tightly. The apples are distinctly visible over Nathan's right shoulder, and it is easier to see the equipment in the foreground. Although the basic transmitter/receivers appear similar to the one pictured a few months earlier in the *St. Louis Post-Dispatch*, Nathan had replaced both the earpiece and the mouthpiece. Moreover, the prominent display of the tray containing 23 dry cell batteries implies that they, not the Stubblefield earth cell, had become the preferred power source.

Within a few weeks, Gerald Fennell visited Nathan at home again. This time he offered, on behalf of the Wireless Telephone Company of America (which he had formed in February under the lenient laws of the Arizona Territory), to pay all expenses for Nathan and Bernard to travel to Philadelphia and New York where they would demonstrate the wireless telephone system for prospective stockholders. Fennell also agreed to handle all the publicity for these events so that Nathan could concentrate on improving his invention. In exchange for his services and the exclusive rights to his invention, the company proposed to give Nathan 500,000 shares of stock, which had the initial price of 25 cents a share. Although no copy of any contract, corporate resolution, or any other legal document survives, Nathan apparently agreed to these terms, and Fennell advanced him $800 for expenses. Within the next few months, several of Nathan's friends and associates in Calloway County bought stock in the company as well.

To prepare for the trip east, Nathan built a new

Stubblefield family at home, April, 1902

Pogue Library

improved model of the wireless telephone. Although it worked on the same natural conduction principle, it was visibly different, more powerful, and capable of greater transmission distance. He and Bernard posed with the equipment alongside the house one day in late spring. The first obvious change was the look of the cabinet. Instead of rough pine, the new model was stained and varnished like a fine piece of furniture. It was noticeably larger at about five feet hall, just slightly taller than Bernard was. The height may have made the controls easier for a grown man to reach, but it also provided more room inside for extra batteries. A new rotary switch allowed the operator to add more power simply by connecting more cells. Nathan wanted to avoid sending out for more batteries as he had done in Washington. There were multiple contacts on the sides for transmission and reception lines to ground. Finally, the earpiece and mouthpiece are larger and perhaps more efficient.

Pogue Library

Nathan and Bernard Stubblefield at home, 1902, with improved wireless telephone equipment for the Philadelphia demonstration

Two other new pieces of equipment emerged. One appears to be a portable receiver, attached to an induction coil about fifteen inches in diameter. The earpiece is very large. The second appears to be a companion transmitter with a long metal horn, typically used as an acoustic amplifier. It is also connected to an induction coil about eight inches in diameter. These were contingent devices, useful on windy days or in other marginal reception conditions.

For his part, Gerald Fennell promoted the Philadelphia demonstrations well. This time Bernard went along. The first one took place on Decoration Day, May 30, and Nathan did several more through June 7, when they packed up and moved on to New York. Nathan was a little alarmed that the company had already been formed and that its publicity mill was running at full tilt. But he had little time for questions, and Fennell had less for answers.

The initial tests were staged in lush, spacious Fairmont Park. Bernard was stationed on the second floor of Belmont Mansion, a mile away. Dignitaries present included: Edwin Houston, one of the founders of General Electric; Albert Crump, general manager of the Philadelphia Key Stone Telephone Company; Henry Clay Fish, treasurer of the Wireless Telephone Company of America; and Prof. Archibald Frederick Collins, a local wireless inventor who had built a system similar to Nathan's. Numerous journalists also attended. One wrote in the *Philadelphia Inquirer*:

> A successful demonstration of wireless telephony took place at Belmont Mansion yesterday. ... All who placed the receiver to their ears went away convinced that the Wireless Telephone Company of America is able to do all it claims. The inventor of the system, Nathan Stubblefield, and his son were present and demonstrated the advantages of the machine. Mr. Stubblefield's son operated the receiver and transmitter on the second floor of the hotel, and Mr. Stubblefield and about forty or fifty persons listened to him in the adjoining woods.

Pogue Library

Nathan and Bernard Stubblefield in Philadelphia, May, 1902, transmitting wireless telephone signals from a room in the Belmont Mansion at Fairmont Park.

Nathan Stubblefield, far right, and a group of witnesses at Belmont Park, Phildelphia, May 30, 1902. Edwin Houston is just to the right of center, holding a telephone receiver to his ear. A. Frederick Collins is second from left.

Others were impressed with the results, but had a difficult time explaining just how the system worked. The *Philadelphia Press* reported:

> Inventor Stubblefield bases his system of wireless telephony upon earth currents of electricity, which, if broken by the vibrations of a diaphragm in a transmitter will produce a corresponding disturbance in the transmitter. His invention consists of a specially constructed system of batteries connected with the receiver powerful enough to produce the disturbance necessary. Seated in an upper room of the Belmont Mansion, Mr. Stubblefield talks into his receiver as one would through an ordinary telephone. A disturbance in the earth's magnetic field results and this disturbance being caught up by a receiver, the auditor several yards distant can hear the voice as distinctly as in a telephone connected with a wire.

The account in the *Philadelphia North American* of May 31 was most intriguing:

> The Wireless Telephone Company of America gave a demonstration of its system to newspaper men and a few invited guests at Belmont Mansion yesterday. Messages were sent with apparent success to a distance of half a mile. In order to demonstrate the new telephonic idea, a public exhibition will be given today at Belmont Mansion, when at least a dozen receivers will be provided.
>
> A message will be sent at a central station and received simultaneously by a half dozen persons. The method of transmission of the wireless telephony is identical with the Marconi system, except that the earth, instead of the atmosphere, is used as the connecting medium. Wires radiate from a transmitting machine, which is built on the lines of an ordinary telephone transmitter. Each wire is connected with a tree, which transmits the sound vibrations to the earth.

> At the receiving stations are two ordinary steel rods, implanted in the earth twenty feet apart, from which wires extend to a telephone receiver. Any message from the transmitter may be heard on the receiver at any distance by merely placing the steel rods in the earth.

Because the text says that this demonstration "will be given," this story is obviously not an eyewitness account. It is unclear how the wires from the transmitter were "connected with a tree." Were the wires slung across a limb to follow the trunk down to the ground rods, or was the tree a substitute for them? No explanation was forthcoming.

At the final demonstration in Philadelphia, Nathan increased his distance to a mile and a half. With the success of the Philadelphia events, the Wireless Telephone Company of America had plenty of propaganda for its prospectus and press releases. It boasted of Nathan's successes with page after page of excerpts from published articles, carefully chosen and edited to make each demonstration appear perfectly successful and unique. The *New York Times* piece on the Washington test was not included, and the one from *Scientific American* was missing the original references to Preece and Collins who had mastered similar technology. Instead, the prospectus was a shrewdly crafted sales pitch that read:

> The Wireless Telephone Company of America has been organized with a capitalization of $5,000,000 for the purpose of developing a new art. It has acquired from Nathan Stubblefield of Murray, Ky., the inventor, all the rights, both foreign and domestic, to the inventions, discoveries, and improvements in wireless telephony and telegraphy.
>
> When the vast savings accomplished in the cost of installation by the new discoveries are taken into cosideration [sic], there is every reason to expect that the magnificent financial record made by the Bell Company's earnings will be

> paralleled. ... The initial offering at the exceptionally low price of 25 cents per share will not occur again, on account of the rapid progress now being made and contracts which the company is securing.

The prospectus went on to explain that the true advantage of the Stubblefield system was eliminating the cost of installing wires in a conventional telephone network. The company also had the rights to the undefined "Stubblefield Uninsulated Wire System," which it claimed would reduce the cost of submarine cables by 90 percent. Nowhere in the document is there mention of the broadcasting capabilities that Nathan found so fascinating. Also, Gerald Fennell's name doesn't appear in the list of principals of the company nor anywhere else in the document.

Plans proceeded for the New York demonstrations. Although Nathan was still having problems with the pace of the business and lingering doubts about the man who got him involved in it, he persevered. The company secured permission to conduct the tests in Battery Park, at the southern tip of Manhattan Island, for 30 days beginning June 11. Almost immediately, Nathan began to experience problems. Bernard said that they never could find a good ground connection because the soil was so rocky. They tried to run a line across the street to a better spot. The police made them take it down because it crossed the streetcar tracks.

They probably encountered a new challenge in that location – poor signal to noise ratio. By 1902, lower Manhattan Island was among the most electrified places on earth. The process began with Edison's Pearl Street generating station in the 1880s. With all the grounded electrical circuits in the vicinity, there had to be an audible hum at 60 Hz nearly everywhere Nathan and Bernard sank ground rods. In many cases, the interference was likely louder than the transmitted sound. Because the system received all electric signals in a 30 - 5 kHz spectrum and could not be tuned to another channel, there was no way to turn the noise off. That was the downside of its broadcasting capability.

Why didn't they move the demonstration elsewhere, like to Central Park where the soil was more favorable and electric connections less concentrated? Perhaps they wanted to use Battery Park because it was the site of Morse's failed submarine cable demonstration 60 years earlier, an incident that stimulated his imagination to invent the first wireless telegraph. Or perhaps the proximity to the harbor provided the chance to do another ship-to-shore transmission. More likely, the location choice was a practical matter related to selling stock. The Battery was the public park closest to Wall Street.

297

INCORPORATED UNDER THE LAWS OF ARIZONA.
CAPITAL $5.000.000.
FULL PAID NON-ASSESSABLE.

-50-

THE WIRELESS TELEPHONE COMPANY
OF AMERICA

This Certifies that Hugh P. Wear is the owner of Fifty Shares of the Capital Stock of THE WIRELESS TELEPHONE COMPANY OF AMERICA. transferable only on the books of the Company in person or by Attorney on surrender of this Certificate.

Witness the Seal of the Company and the signatures of the President and Secretary.

Registered the 1st day of Aug 1902
Trust Company of the Republic
REGISTRAR.
by
SECRETARY

VICE-PRESIDENT

SECRETARY

Pogue Library

Stock Certificate, Wireless Telephone Company of America, issued to Hugh P. Wear of Murray

For whatever reason, Nathan's system failed to work. The company was upset. Fennell even went so far as to suggest that Nathan and Bernard bury some wires to surreptitiously connect transmitter and receiver. That way, the demonstration was sure to work. Nathan refused and became even more suspicious of the company. So far, he had seen Fennell and his associates do nothing but issue press releases and sell stock, some of which his friends back in Murray owned. From his point of view, it seemed that no one at the Wireless Telephone Company of America was interested in developing his technology and selling telephone systems. Nathan had never seen his stock certificates. The Trust Company of the Republic held them, perhaps as collateral. And Fennell had applied for a life insurance policy on Nathan, with the company as beneficiary, without his knowledge or permission, a standard procedure in corporate America but a concept foreign to the West Kentucky farmer. When Fennell asked him to sign the application, Nathan decided he had had enough.

On June 19, Nathan wrote to S.N. Turner, Secretary of the Wireless Telephone Company of America:

> Mr. Gerald Fennell the promoter of our Company has had a letter from me of which you have a copy. He has answered same cleverly evading and practicing fraud or deception as usual, and there remains nothing for me to do but to go home. I regret very much such has been the ending and regret very much that my name is connected in any way with the concern and shall take immediate steps when I reach home to turn on the lights that the public may not be swindled by this fellow as I have been. It becomes my duty (as I am one of the directors) to see that this be a fair legitimate business or I am a party to the fraud that may be committed. I very respectfully decline having a thing to do with the business until [it] is in every way put on an honorable basis and put in the hands of men who will conduct same. (The letter referred to is of date of 17th of June, 1902, and is in nature a complaint against the aforesaid promoter and should be seen by every member of the Company.) If you Sir depend on this man Fennell to put the matter before the Company they will never know the facts as I have presented. I therefore ask that you provide each of them with copy of same, that they may have a chance to adjust the matter after which time should they fail to act then it remains to be clearly seen that they are parties to the swindle of me out of my inventions and the

defrauding of the public. I shall notify each and all of them that you have such document in your possession. To comply with my duty since I have signed over everything that I have to do with this concern today with Mr. Walter Hood, your fellow associate clerk as witness, I turn over to you all the property in my in my possession belonging to the Company and depart for my Kentucky home with a feeling of gratitude for some New York people who with me have watched the steps of this man Fennell through many hours of uneasiness to me.

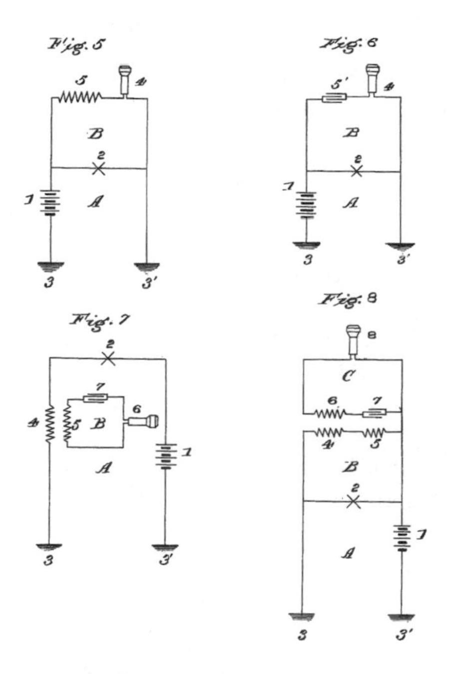

Wireless Telephone, A. Frederick Collins, patented 1905. While this bears some resemblance to the Stubblefield natural conduction system, neither Nathan nor the Wireless Telephone Company of America ever applied for a patent.

So Nathan and Bernard left for Murray, leaving behind his most promising invention for which he could no longer claim ownership. He had nothing else to do with the Wireless Telephone Company of America, nor it with him.

The company, however, survived a little longer by securing the services of A. Frederick Collins, the inventor who witnessed Nathan's first Philadelphia demonstration. It is altogether possible that this deal was already in the works and that Fennell's actions in New York were designed to provoke Nathan to take the action he did. Within a month, Collins' name appeared in the company prospectus in place of Nathan's, complete with a reprint of an *Electrical World* article by Collins about wireless telephones.

For his part, Collins demanded more money, but was also more valuable to the company. He knew more about wireless than Nathan did and had wider contacts in the scientific and professional engineering worlds. In his article "The Collins Wireless Telephone System" for *Scientific American* of July 19, 1902, Collins describes a system very much like the earlier Dolbear patent, complete with an illustration. Although it includes all the basic elements of a true radio system and is designed to transmit and receive high frequency Hertzian waves, the design has several flaws. For example, it lacks an efficient electromagnetic wave detector in the receiver. Collins admitted that it did not work at distances greater than three miles, and most of the signal probably passed through the earth. In passing, he described the other wireless methods, commenting that with the natural conduction technology that Nathan and others used, "a very few quantitative tests will show that the limitations appear almost before its commercial value begins." He eventually got a US patent in 1906 on his wireless telephone system. Although no detailed plans for Nathan's 1902 model exist, the photographs indicate obvious dissimilarities to the Collins design. Collins uses elevated condensers as antennae, while Stubblefield incorporates no aerial devices. Collins may have stolen ideas from other inventors, but apparently not from Nathan.

The Wireless Telephone Company of America fared little better with Collins. It quietly ceased operations, probably by 1903. In his 1905 book *Wireless Telegraphy: Its History, Theory, and Practice*, Collins mentioned no connection with the company or with Nathan Stubblefield. In 1907, he formed his own company to exploit his invention, then merged with two other companies to form the Continental Wireless Telephone Company. Collins and his partners sold more than a million dollars in stock before the United States Attorney of New York indicted them for mail fraud. Convicted in 1913, Collins plead that he had devoted all his time to the invention and knew little of how his partners conducted the business. The judge was not impressed, saying that "there was no doubt that if he had invented a workable apparatus before advertising he would have been acquitted, but that, as it was, he had made false representations as to his work and what he had accomplished." He sentenced Collins to three years in prison.

Nathan was better off getting out when he did.

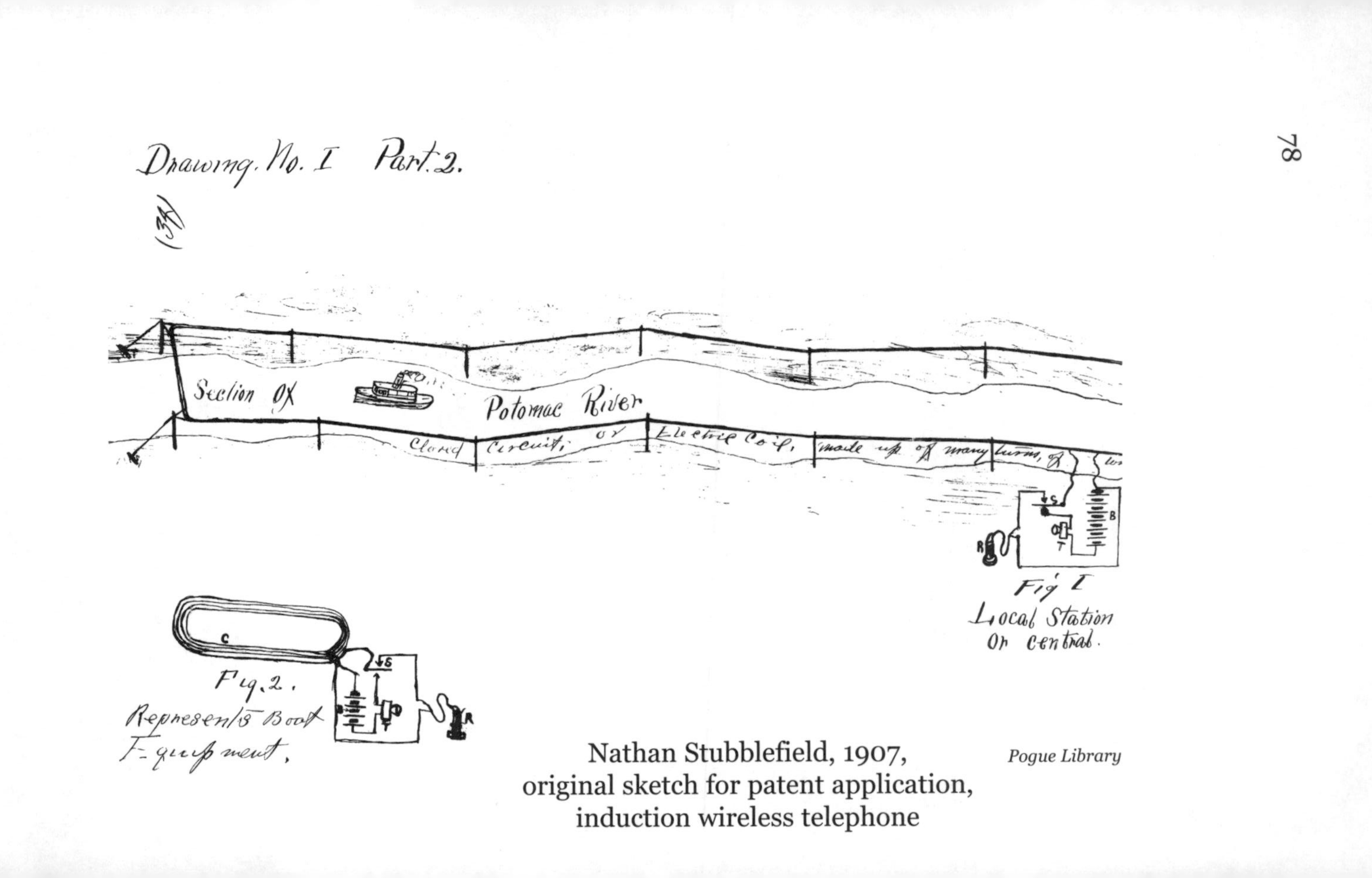

Nathan Stubblefield, 1907,
original sketch for patent application,
induction wireless telephone

Pogue Library

Back on the Farm

When Nathan departed New York in June 1902 and left behind his invention, equipment, and documents related to them, he also ruled out any further work in natural conduction wireless telephony. Although Nathan never patented the device nor received anything but worthless stock from the Wireless Telephone Company of America, he was nonetheless under a contract that granted the company ownership of his invention. He had received at least $800 compensation for his work on the demonstrations. In his mind, the scoundrels had won this round.

He had two choices. He could devote his full effort to his family and farm, or he could revert to his original work with induction wireless telephones that he undertook before his agreement with the Wireless Telephone Company of America. Still convinced that he was on the verge of a technological breakthrough that would make him wealthy, Nathan chose the second path.

Never one to encourage visitors, Nathan became even more secretive in this phase. He felt that he had been compromised in his early wired telephone business and again by the entrepreneurs from New York. That was enough. Yet he also began to keep written records of his experiments, perhaps in preparation for a patent application. Bernard was now an experienced assistant. Nathan soon added other members of the family to the project, mainly as witnesses. His wife Ada and children Bernard, Pattie, and Victoria signed the following affidavit:

> In order to establish date of New Invention in Wireless Telephony of Nathan B. Stubblefield for any future needs that might arise technical description herewith provided of apparatus used.
>
> At home Jan. 15, 1903
> This day Nathan Stubblefield transmitted wireless telephone messages one hundred and twenty five yards without ground connection his latest development in wireless telephony. This affidavit is the first documented message trans-

mitted by this system through sixty yards space. ... This message was transmitted at 8 o'clock night of Jan. 15 by Bernard Stubblefield and received by Nathan B. Stubblefield the inventor and received again by the below signed as witness.

Pattie L. Stubblefield

Pogue Library

Nathan and Bernard Stubblefield in the workshop at the Stubblefield farm, Murray, 1902

Since this device worked without a ground connection, it was totally dissimilar to his wireless telephone of 1902. Another affidavit, signed by Pattie and Victoria a year later, gave further details:

At home Jan. 23 of 1904
This is to certify that we the undersigned date above shown heard at a distance (roughly stepped) of six hundred feet harp music by Wireless Telephone, Nathan Stubblefield's secret invention where in no earth connection is used described as follows and understood by us.

Circular coils of No. 28 magnet wire 26 ft. in diameter with forty convolutions with forty eight cell dry batteries connected in with coil and carbon ball transmitter, as transmitter of messages

Receiver as follows two coils wire seven feet in diameter containing 33 convolutions each. First coil office wire No. 20 second or top coil of No. 20 Magnet wire with two bell receivers.

It is not understood by us or father whether it is by electromagnetic wave that this is done but well known that simply a primary current passes through coil and transmitter connected one to each distinct circuit or coil. Bernard B. Stubblefield transmitted music from coil just west of house, our home, to forked red oak tree on land east of our house, with its forks pointing north and south with poison ivy growing on its west side a snag of a tree with knot near top rather on the south side.

Given our hand this Sunday night Jan. 23 all with a view of establishing facts as they exhist [sic] for the future interest of Nathan Stubblefield the inventor and our father who was with us in this test.

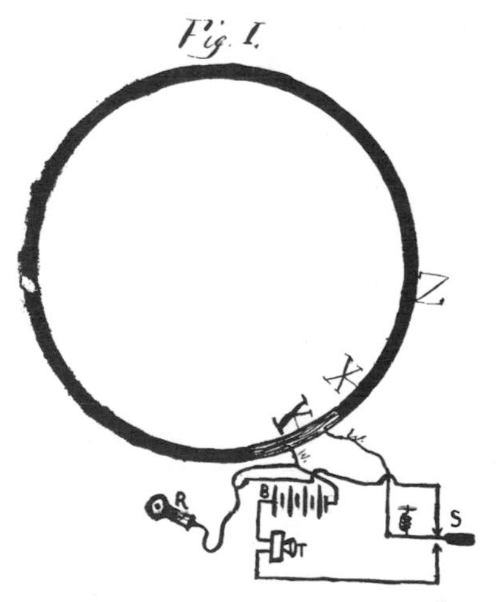

Pogue Library

Bernard Stubblefield's sketch of a coil for the induction wireless telephone, 1907. Nathan's notes said that the coil was 60 feet in diameter and transmitted and received "one fourth mile nicely."

The final two affidavits came two weeks later, on February 4. The first, signed by Bernard, Pattie, and Victoria, reads:

> We the undersigned testify to the fact that this day a coil of No. 20 copper wire, the coil forty feet in diameter with 42 convolutions, with 48 cells of dry battery and a mycrophone [sic] transmitter was used in transmitting wireless telephone messages — conversation and harp music four hundred and twenty three yards from our residence with no sort of earth connection. A coil as receiver of 26 ft. in diameter of No. 28 magnet wire with 40 convolutions with a double pole receiver but no sort of earth connection. Other station lying westward in a woods from the home place located by a dogwood tree of small size known to us.

The second, also signed by the three children contains a note from Nathan at the end:

> This is to certify that we the undersigned did this day receive wireless telephone messages, conversation and harp music four hundred and twenty-three yards distant from the transmitting station <u>without any sort of earth connection</u> by means of Nathan Stubblefield's new system of Wireless Telephony, claimed by him to be done through the Hertzian or electromagnetic wave process and practical for great distances, either stationary or portable.
>
> Note: The above are sons and daughters of mine who understand the technical features of my inventions.

There are some interesting revelations in these documents, all of which appear to be in Nathan's handwriting. First, as noted above, Nathan went out of his way to stress that he used no earth connection, thus distancing himself from the 1902 invention and any potential patent interference action. There is no mention of any broadcasting

Stubblefield and family at the farm, 1907, with induction wireless telephone system. Nathan is third from left, holding a portable receiver. The large transmitter coil is behind them, in the center.

Pogue Library

capability for this wireless telephone. Clearly, the intent for it is two-way communication. Another goal seems to be a portable receiver, but the range is not very promising. The final experiment uses more than a mile and a half of wire in huge coils to send a signal about one third of a mile, approximately the limit that John Trowbridge claimed for this technology in 1891. Under the circumstances, a wired telephone would be more cost effective. Finally, Nathan appears to be confused about electromagnetic waves. In his subsequent patent application, he made no claims about Hertzian wave transmission and did not submit a design for a receiver capable of detecting them.

Realizing that transmission distance would be short, especially if he minimized coil diameter to build small portable telephones, Nathan decided to concentrate on wireless communication with moving vehicles including riverboats, trains, stagecoaches, and later automobiles. Using the riverbanks, railroad tracks, or roads as a guide, he could locate the primary coil on poles above the path and surround it with a magnetic field. The secondary coil, on the roof of the vehicle, could then remain parallel to and just a few feet away from the primary. But this design ran afoul of the US Patent Office. It pointed out that Lucius Phelps and others received patents on similar inventions more than 20 years earlier.

Moreover, the external forces of technological change worked against Nathan's efforts. Whereas his mentors were Morse, Henry, Bell, and Edison, there was a new generation of electricians at work on the challenge of wireless communications. They took their cues from Hertz, Lodge, Marconi, and Tesla. Many of Nathan's competitors in this global market had university educations, financial backing, and extensive laboratories and test facilities. While Nathan tried to make 1880s technology work better, they had already tossed that aside to pursue wireless by electromagnetic waves. Nathan was struggling to makc his induction wireless telephone transmit a mile. Meanwhile, Marconi used Hertzian waves to establish transatlantic wireless telegraph service. Electricians like Lee de Forest, Reginald Fessenden, and John Stone Stone were rapidly

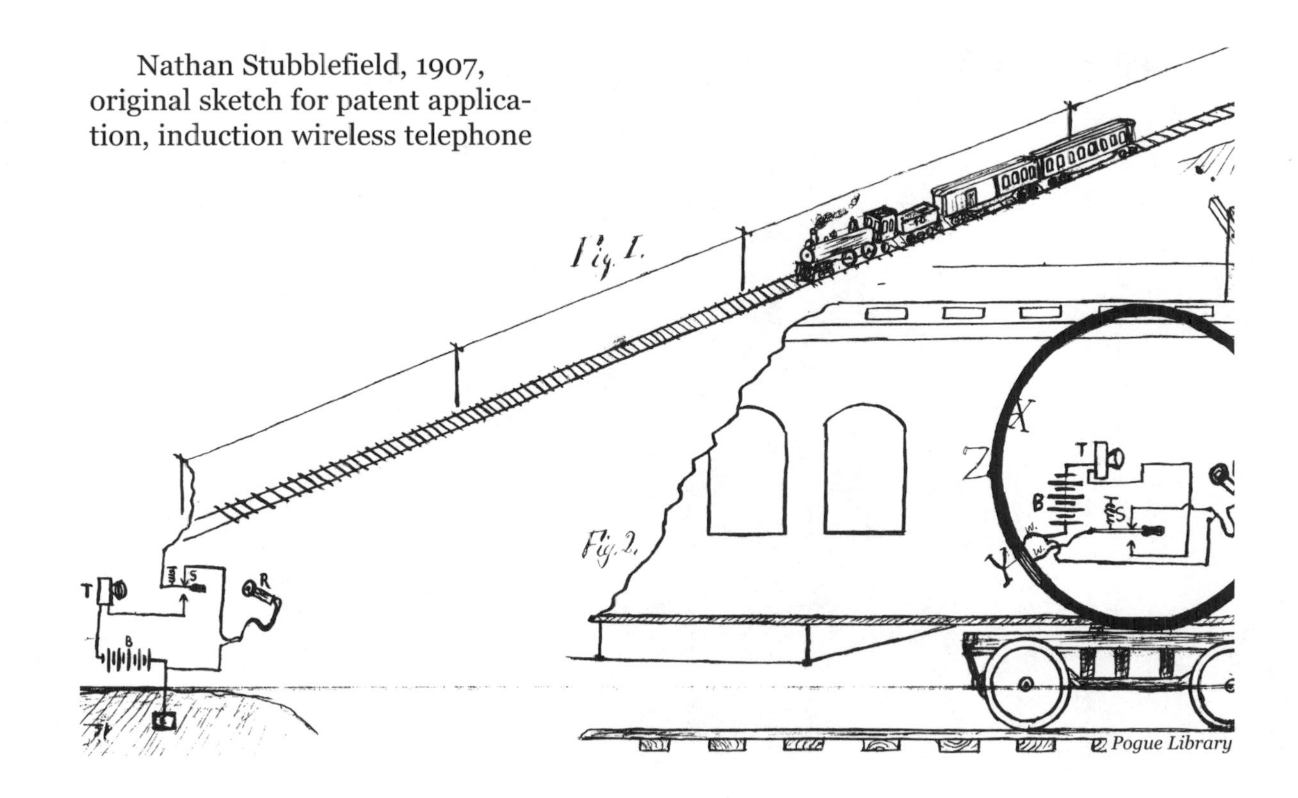

Nathan Stubblefield, 1907, original sketch for patent application, induction wireless telephone

making improvements to that technology so that, by 1906, it would be used for sound transmission and eventually become what we now call radio.

Nathan stubbornly pursued his goal of getting a patent on the induction wireless telephone. He first contacted a patent attorney in 1902 and received an opinion that he could not get a patent on the device, presumably because it wouldn't work. After successful experiments in 1903 and 1904, Nathan tackled his next problem, raising funds. By late 1906, he had convinced six men from Murray to back his enterprise with at least enough capital to build a working model, have photographs made and promotional brochures printed, and secure a patent. Nathan then contacted Washington patent attorney E.L Siggers, with whom he had dealt in the past. After his initial reluctance, Siggers agreed to pursue the patent in Nathan's behalf for an initial sum of $100.

Upon examining Nathan's original application, Siggers wrote on February 25, 1907:

> My attention has been called to certain patents issued in 1886, which bear a very close relation to your invention, even if they do not anticipate it. These patents have expired and are now public property, but the fact that they have expired does not justify the Patent Office in issuing a patent to you on the same or an equivalent invention. I hesitate in proceeding with the application.

Negative opinion to the contrary, Nathan replied with details of the numerous differences between his design and the earlier patent, issued to Lucius Phelps and described in Chapter 4. The crux of this argument was that he used a coil of wire to increase the induction field strength, while Phelps had only a single wire. Although the Patent Examiner did not refer to the Woods induction wireless telegraph, that device did utilize a coil for more power. Nathan also stated that this was his first knowledge of the prior work and that it appeared to be impractical.

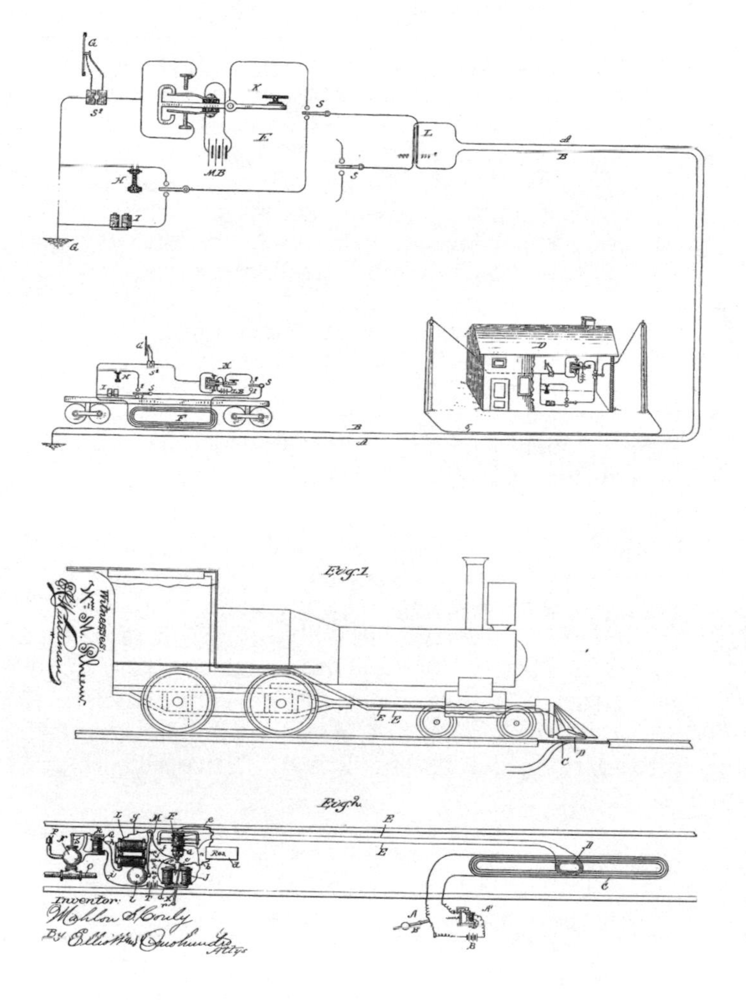

Induction wireless communication systems patented by Lucius Phelps, top, 1886, and by Mahlon Conley, 1894, bottom. The US Patent Office said that Nathan utilized technology already protected by these prior patents.

As predicted, the initial reply from the patent examiner in April, 1907 was unfavorable:

> The Examiner questions whether there is any advantage as alleged in the plurality of convolutions in applicant's longer conductor. The resistance of the circuit of which it is a part must lie wholly in the conductor itself, hence the resistance must be as much greater as the wire is longer, and it follows at once that the conductive material were thrown into a single conductor.
>
> The claims are rejected on Phelps 334,186, Jan. 12, 1886, Class 179, Subclass 82, in connection with Conly with 528,122 of Signaling Railway Cab, in view of which there would be no invention in multiplying the convolution of Phelps's conductor.

Nathan was in a frustrating quandary. In order to secure a patent, he had to dispense with the argument that his idea was unique and original and, instead, advance one that his invention represented substantial improvement on the work of others. Yet it was his claim to originality that had attracted the few financial backers on whom he relied to solicit new investors. He sent Siggers a lengthy and spirited rejoinder to the Patent Office rejection, in which he stated:

> Now if there is any difference in a single convolution (in electrical action) and a multiple of, or many convolutions, there is certainly a difference and a great one, between what Felps [sic] has a patent on, and my inventions, which show marvelous improvement, which on its face constitutes real, true invention, as is in keeping and has been for ages, in the rulings of the Patent Office.
>
> As to the Conly patent #528,122, Oct. 23, 1894 Signaling Railway Cab — There isn't the remotest kinship to my inventions. He was not the inventor of the induction coil "the father of his thought." He hasn't a corner on induction appa-

ratus, only in his improvements, there exists between his and mine no kinship on earth! If so, how could he have gotten a patent over Phelps? Havn't [sic] I a corner on somebody?

In conclusion allow me to state that I am willing, ready, and anxious, on short notice to make a demonstration at my own expense, to prove up, and ask that I be granted the chance due a man, who has been an experimenter on Electrical Science for most sixteen years.

I offer as reference to my reliability of my statements the <u>men</u> of Calloway County Kentucky, of whom six of the brainiest, and best, have put up money on my ability as technician (excuse immodesty) namely Mr. Conn Linn, B.F. Schroader, Jeff D. Rowlett, R. Downs, Geo. C. McLarin, J.P. McElrath.

The Financial Supporters of this enterprise, all of Murray, Kentucky, are Senator Conn Linn, Mr. B. F. Scroader, Mr. R. Downs, Mr. J. D. Roulett, Mr. Geo. C. McLarin, Mr. John P. McElrath.

These gentlemen believed in me, and my invention. We have been granted an allowance of the United States patent, and by our attorneys at Washington, guaranteed the issue of patents in the following Foreign countries; Canada, England, France, Spain, and Belgium, and it is our aim to apply for five additional Foreign patents, before the U. S. patent is made public, through or by some plans, presently to be cited on next page, under head of *Private Prospectus.*

Pardon Immodesty — But a Word about Ourself

The author of this invention, Nathan B. Stubblefield of Murray, Kentucky, the pioneer electrician of that town, has been an experimenter in electrical science for many years; was the author and patentee of *The Stubblefield Acoustic-Telephone* nineteen years ago; eight years later the inventor and patentee of an electrical battery, patented in the United States, England, and Canada, which battery invention, proved the foundation of the present inventions in Wireless Telephony. An experimental place just West of the town of Murray, is the home of the *extended research* necessary in the working out of problems. Here, aided by inteligent effort and eternal push, with the assistance of a son, Bernard B. Stubbllfield (now twenty years of age), my inventions, which I trust will be worth sometning to the world as well as myself, have been made.

Pogue Library

Stubblefield's 1908 Prospectus

Nathan decided to move to Washington. He reasoned that he could speed up the patent process by being able to communicate with Siggers in person and being available for his proposed demonstration before the Patent Examiner. He would also be distant from his partners who may have had second thoughts about their investments following the Patent Examiner's rejection. Bernard joined him but had to take a job to help pay expenses. To pass the time, Nathan tried to drum up money. In early December, he mailed a prospectus to fourteen potential investors in Murray, and followed with a somewhat desperate letter to his original "big six:"

> I am pretty sure that out of this 14, you people can enlist ten of them. These people named I think have confidence in my ability to turn money out of this business, and if you people will use your influence the business will move right on. I am making acquaintances here and am confident of success. Let me hear from you soon as is convenient, some sort of expression.

Pogue Library

Nathan Stubblefield, 1908, with the portable receiver, or desk set, for his induction wireless telephone system.

Siggers' arguments on Nathan's behalf prevailed eventually, and on May 12, 1908, more than a year after the application and initial rejection, Nathan received US Patent 887, 357 for a Wireless Telephone. The patent stated:

> I, Nathan B. Stubblefield, a citizen of the United States, residing in Murray, in the county of Calloway and the state of Kentucky, have invented a new and useful Wireless Telephone, of which the following is a specification.
>
> The present invention relates to means for electrically transmitting signals from one point to another without the use of connecting wires, and more particularly comprehending means for securing telephonic communication between moving vehicles and way stations. The principal object of this invention is to provide simple and practical means of a novel nature whereby clear and audible communication can be established, said means being simple and of a character that will permit certain station mechanisms to be small and compact.

The specifications mention that with a large primary transmitting coil, "the other may be very small and speech or sounds can be transmitted comparatively great distances one to the other." In the diagrams, however, these distances never exceed 100 feet. Upon receiving the patent, one of Nathan's first actions was having a professional photographer take his picture with a model of his desk telephone for future publicity. It is quite possible that this was the only model that existed.

Despite grandiose plans for an "Ideal System of Wireless Telephony" that would put "the homes of Europe and America in communication," there is no evidence that Nathan ever sold a single installation based on the 1908 patent. The technology was obsolete and impractical. Although he was heralded in the book *Our Wonderful Progress* as the inventor of the wireless telephone, the local newspapers rarely mentioned Nathan or his invention.

His partners considered legal action for fraud but demurred because they were unlikely to get any compensation. Nathan was broke. His wireless telephone had been stillborn. He didn't even own the farm that he lived on. It was the end of his career as an electrician.

Hard Luck and Trouble

Although Nathan retained the dream of earning a fortune from his electrical inventions, the pressing urgency of his family's financial situation brought him down to earth. He decided to channel his scientific interests into a new and hopefully profitable endeavor, an industrial school at his home. He named the institution Telephondelgreen and boasted that it "turns everything to gold it touches. Great institution it."

Since Nathan had been schoolmaster for his children, his role was familiar. He came up with the idea in late 1907, during his stay in Washington, and was ready for business the next fall. On August 28, 1908, he proudly drafted "Our Course of Study" in his own flourishing hand. The curriculum was unique in several respects. It included penmanship (Nathan's specialty) and standard grammar school subjects like reading, geography, civics, history, and composition. Nathan expanded the program to encompass single-entry bookkeeping, "Easop's [sic] Fables, "Bible Reading First for 15 Years," and instruction in etiquette and hygiene. He offered "Agriculture and Fruit Growing ... in the thorough and bread-winning sense." There was substantial emphasis on science and technology, from chemistry, physiology, and natural science to a detailed "Technical Course." The reading list contained books by Houston, Dolbear, Edison, and other authors in Nathan's personal library as well as "Current Literature" like:

> *Kind Words, Child's Gem, Practical Farmer, Courier Journal, Success Magazine, Woman's Home Companion, Bullitin [sic] of Progress, Telegrapher's Journal, Electrical World.* The representative technical journal of America and many good books on various topics and in line with Sound Ethics in the light that I have viewed the proposition of Life and How To Live It.

Like the wireless telephone ventures before it, Telephondelgreen was a failure. There is no evidence that

To Whom This May Concern
Thinking it might be of general interest to know of about what is taught in our Home School I thus hand you Our Course Of Study.
This Aug 28, 1908
And Am Fifty
N. B. Stubblefield

Bible Reading First for 15 Years.

Agriculture and Fruit Growing. is taught, in a thourough bread winning sense.

Literary Course
Spelling—
Reading.
Arithmetic.
Geography.
History.
Physiology
Hygiene.
Composition and Rhetoric
Civil Government
Chemistry
Natural Science.
Book-keeping
Penmanship
Art.
Etiquette,
"Eesop's Fables."
(Columbian Text Book. English and Grammar. Single Entry and)

Technical Course.
"Magnetism." By Houston and Kennelly.
"Experimental Science." By Geo M Hopkins.
"Dynamo and Motor Building" By Parkhurst.
"Telegraphy." By Theo Edson and Valentine Bros.
"Telephony." By Lockwood and Dolbear.
"Electrical Measurements." By Edwin J. Houston.
Electrical Dictionary. By Houston, of several thousand Words, terms and Phrases.
With other works in Electrical Science.

Current Liturature.
"Kind Words."
"Childs Gem."
"Practical Farmer."
"Courier Journal."
"Success Magazine."
Womans Home Companion
"Bulletin Of Progress."
Telegraphers Journal
"Electrical World." The representative, technical Journal of America, and many good books, on various topics, and in line with Sound Ethics, in the light that I have viewed the proposition, of "Life and How To Live It."

Wrather Museum

Curriculum for Stubblefield's school, 1908

Stubblefield's home school,
Telephondelgreen, 1907

Wrather Museum

Nathan had any students other than his own children. His secretive, almost reclusive, lifestyle probably doomed the school from the start. Why would anyone send their children to study under a man who had his farm bugged with wires and bells to warn him of intruders and who threatened to chase off visitors with a shotgun? In January 1911 he wrote to a friend, "after this winter this school will close and the children will be put in school here or elsewhere."

Nathan and family also faced a mounting debt burden that threatened to make a difficult life intolerable. In March 1911, he went over his unpaid bills and IOU's, some dating back many years, and wrote:

> It is yet my aim as before decided to pay all these old debts with interest if I can make money in a lump out of some of my inventions, that is make thousands. Nothing would make me happier than to be able to pay these old debts after so much pervation [sic] and persecution has come to me. I am, as you can see, trying to face myself in my own estimation, that I am honest and will do the right thing possible in any reasonable degree of reason. Five hundred thousands of shares of wireless telephone stock, the certificate of deposit will be seen framed as evidence that I have been trying to pay and if this stock had ever got to parvalue I could have paid. Now the old paper and this are evidence that I wish to keep in memory those matters for my childrens [sic] sake and that I may know where the conscious fund must go if I do make money yet. I am now fifty years old and notwithstanding that the life has nearly been worried out of me in bringing up of this family and in worring [sic] with old creditors and new ones and the whole hell bent of night riders and damn fools. Generally I am going yet some day to answer this question, did he ever pay them old debts.

Sadly, in 1911 Nathan also received his highest praise in years in the book *World's Work*. In a chapter titled "The Wireless Telephone," the text reads:

> Now comes the announcement that an American inventor, unheralded and modest, has carried out successful experiments in telephoning and is able to transmit speech great distances without wires. The inventor is Nathan Stubblefield. The first public test of telephoning without wires was made at the Kentucky village where the inventor lived on the first day of January, 1902, only a few weeks after Marconi's success in signaling across the Atlantic by telegraph without wires.

The piece continues with a lengthy quotation from the 1902 *St. Louis Post Dispatch* article and from a Frederick Collins article on wireless telephony. There is no mention of Nathan's subsequent work with induction wireless, or of his 1908 patent. The final paragraph, however, refers to the contemporary advances in the art:

> The United States government has equipped many of its war vessels with the wireless telephone apparatus for communication at short distances, say from ten to fifteen miles, and under favorable conditions, it has been found to work well. As a general proposition, however, it has no particular advantages over the wireless telegraph, aside from the fact that the voice of the speaker may be heard and identified.

At the time, the US Navy was testing wireless telephone equipment from de Forest, Fessenden and others, all of which used electromagnetic waves. There was no interest in induction wireless systems like Nathan's. On the other hand, the US Army Signal Corps continued to utilize natural conduction wireless telegraph equipment through World War I. These portable devices were practical communication tools for artillery spotters.

In early 1912, Nathan scraped up enough money to return to Washington and try his fortunes there. He landed a job as a bookkeeper, met some distant relatives whom he described in a letter to his brother George:

> These people are kind, proud, and honorable, they have had me to break bread with then, and say they are proud of my scientific accomplishments, but I must say, that at this time, and for a time, my electrical business is laid on the shelf, because of some litigation and a "wee little" pow wow that was pulled off there in Calloway 3 or 4 months ago of which I will not speak further, only to say that it crippled me financially. I am keeping books at the present. In fine health, and getting on pretty well for a chap 50 years of age. And my ambition is not gone either. ... P.S. May add, that my place of business in on Conn. Ave., the swell section of Wash. The finest Resident street of any city in the world.

The litigation that Nathan mentioned was probably the one contemplated by his partners in the 1908 wireless telephone venture. Their actions and his desire to make good on his debts may have forced him to seek employment in Washington in the first place. But soon, Nathan was back in Murray and involved in a more serious and ultimately devastating legal issue.

Living conditions on the farm were insufferable. Oliver, the youngest son, ran away from home in 1912, at age 15. Tired of living in abject poverty and eager to start lives of their own, the three oldest children, Bernard, Pattie, and Victoria, had reached legal age and decided to divide the farm into three equal sections and sell it. Under the terms of William Stubblefield's (their uncle's) will, they could do this. Bernard explained the reasoning to Tom Morgan:

> Uncle Billy specifically willed it to us. He [Nathan] wanted to take it. He'd blowed it in no time if he had got it. Uncle Billy, his brother, he helped him out in giving that place to us. He deeded it to us. He wouldn't deed it to him because he knew that he was sort of like Thomas A. Edison. They told it on Thomas A. Edison that if he got $50,000 he come right in and spent it experimenting. Well, that was the old man — the same way.

In November 1913, Nathan sued his children to stop the sale. He claimed that his brother meant all along for him to have a life interest in the property and a place to live. With this understanding, Nathan included the single acre and the 3-room house on it, which he and Ada owned, in the deed. The judge allowed the sale but ordered that $400 of the total go to Nathan and Ada as compensation for their original property. Nathan accepted the settlement and vacated the property.

It is unclear where the remaining family lived for the next few years. Nathan Franklin left home soon after his 18th birthday, and Helen, the youngest child, married and moved to Tennessee in 1917. Then Ada did what she had promised for many years; she moved back to Paducah. Nathan was homeless and alone.

Always secretive and reclusive, Nathan became a virtual hermit, moving from place to place during the last decade of his life. Details of this period are scant and mainly anecdotal. In 1918, he tried unsuccessfully to land a job with the US Department of Agriculture. He also wrote to the US War Department offering to give his inventions to the government in exchange for public assistance to develop them. After he received a noncommittal response, he never followed through on this effort.

For a time, he lived in the Pottertown community, just east of Murray. Around 1921, Nathan moved onto land owned by Carl Crisp in the Almo community a few miles north of town. One of his first domiciles during this period was a crude structure that he built on Crisp's farm. A neighbor described it:

> Stubblefield picked out four trees about 8 or 10 feet apart, and walled them up by driving poles between them. He chinked the walls with cornstalks and mud. Stubblefield left the floor dirt and that's where he lived and worked.

His cousin Vernon Stubblefield, who often brought Nathan food, was sickened by his living conditions, as he later told Tom Morgan:

> We came to a little knoll there with some oak trees and a nice bank, grassy spot, and right over here to the left was the cabin, and he asked us to sit down, and we talked a while. [We] thought that he would ask us to come up to the house, but he never did. I walked by the front with an eye on my ear, you might say, because I didn't want him to know that I was looking, inspecting anything of his, and it was just very sparsely — I don't see how a man lived there. It would break your heart. Good Lord! I tell you, to see the man dressed up like he was in Washington and the man that I saw that day! Damn, I think he had on a sackcloth apron. Oh me, it just beggars description.

During the winter of 1923, Crisp convinced Nathan to move into an abandoned tenant house, partially dug into a hillside on the farm. It was there that the inventor spent his last years. His landlord also let him do light work around the place in exchange for fresh milk, meat, and vegetables from the garden. Otherwise, Vernon Stubblefield, a few other remaining friends, and his neighbors brought him supplies from time to time. Nathan kept track of most of these gifts with the intention to repay the charity when he finally struck it rich. As grateful as he was, though, he rarely invited anyone into his home. Perhaps he was ashamed of his living conditions, but he claimed that he was still working on inventions that demanded tight security.

Often, Nathan passed on photographs and mementos of the family and his career to Vernon as payment or collateral for his IOU's. Among these items was the article from *Electrical World* of 1898 about Nathan's earth cell, his first national publicity. In the margin, Nathan scribbled on September 20, 1922 a reference to how his battery experiments taught him how to use natural conduction for wireless telephony:

> My first Electrical and Technical Book, patented in the US, England, and Canada. This has con-

> tributed quite a lot to the world's sum of knowledge. It was the bedrock of all my scientific research in raidio [sic] today.

Another gift, the original Canadian Patent for the wireless telephone, carried an interesting notation also dated September 20, 1922:

> Vernon,
> I am the author of two miniature books in the US, England, and Canada, and since you have developed into a radio bug or fan I am sure this will contribute to your efficiency, and if so I should be happy.

These two notes, late in his life, are the only times that Nathan claimed that his inventions had anything to do with radio. They also suggest that his cousin Vernon had as much interest in making the connection as Nathan did.

It's been said that, if you live in a small town and forget what you did last week, just ask anyone and they'll be glad to tell you. As furtive as Nathan became during this period, he exhibited a mystique that was an intriguing subject of local gossip. He may have encouraged this. Within his character still lurked the prankster and the promoter in a Derby hat. Again we encounter the paradox of a man who craved privacy but needed attention to validate his genius.

Reduced to recycling what little wire, metal, telephone parts, and other material he had saved and having only a stack of aged mildewed books and technical journals as resources, Nathan tried to work on his inventions. A few of the earth cell batteries connected in series could supply low voltage electrical power, enough to operate scaled-down models of either wireless telephone system. He apparently tested these devices by concealing receivers in his neighbors' fields. Nathan then surprised them with a vocal warning that their cows had broken through the fence again, or a similar admonition that was as much a practical joke as a scientific achievement.

He also claimed to be working with electric lights, perhaps low wattage bulbs powered by the earth cells. In

1962, Mrs. L.E. Owens, one of Nathan's neighbors, told the *Paducah Sun* on a conversation with Nathan a few weeks before his death:

> He was radiant. He told me, "I've taken light from the air and earth as I did sound." He said he could have the whole hillside blossom with light at the wave of a hand. Before we left he told me, "There are those who would steal my invention — foolish men. Without me it is useless."

The rumors about the lights persisted, but research to date has produced no eyewitnesses to this phenomenon. Other neighbors spoke of how Nathan's shack remained warm in the dead of winter without a fire on the hearth, leading to the conjecture that he had somehow "bottled sunlight." People were willing to believe almost anything about this crazy old hermit, but most of what fueled the rumor mill was pure unsubstantiated speculation.

Late in life, Nathan promised his cousin Vernon that they "will yet add luster to the Stubblefield name." On March 3, 1928, he wrote to his daughter Pattie:

> I never go anywhere, just work like h—- all the time. ... I am plucky as the devil for a man nearly 70 years old. Mind strong and clear. I am not dissipated in any way under the sun, and am just 100 years ahead of my fellows and encased in a mental atmosphere that they cannot penetrate. Have sense enough to hold both ends of the string in my scientific work. ... If I live, my papers will be guilt [sic] edged. Don't be offended now.

Less than four weeks later, Nathan died of malnutrition. He had told a neighbor, Obid Daniel, that he was feeling poorly and asked him to knock on the door if he didn't see smoke from the chimney for a day or so. Daniel did as asked and, on March 30, discovered Nathan's body lying on the floor. He summoned the coroner, J.H. Churchill, who determined that Nathan had been dead about two days. Churchill's son Ronald was present and recalled the grim scene years later in a letter to L.J. Hortin:

> My father ordered me to open the door which was fastened from the inside by a small wire over a nail. It was a small buggy shed boxed-in on the one-room house with a little ramp from the dirt floor to the door opening to the one room. Mr. Stubblefield's body was lying on this ramp with his head stretched back and inside the main room. I can see him now and the impression I had was that he had come that far, maybe for some fresh air. He definitely died of malnutrition as there was only an empty milk bottle and some dried beans in the room
>
> I was the first one to touch his body as all the others were just looking in and I was a couple of steps ahead of my father. There was a large cat inside the room sitting near Mr. Stubblefield's head. It ran when I got closer. This cat had been licking the eyeballs, or sockets, as the eyeballs were gone, destroyed by the cat, I am sure.

They buried Nathan in an unmarked grave in the Bowman family cemetery, outside the fenced area reserved for his noteworthy ancestors.

Nathan died without a will. Bernard received the con-

Pogue Library

Unidentified experimental device, 1928, found in Nathan Stubblefield's house after his death and later dismantled by Bernard Stubblefield

tents of the house, mostly papers, odd coils of wire, and junk. Years later, an archeological survey of the site uncovered wires in the trees and coils, probably the remains of earth cell batteries, buried in the ground. Bernard also got his father's trunk that had transported the wireless telephone gear to Washington, Philadelphia, and New York in 1902. And there was a large black control panel. To this day, no one has been able to explain what it was or for what purpose Nathan used it.

The *New York Times* carried the following obituary:

RADIO PIONEER DIES, POOR AND EMBITTERED

Kentucky Hermit, Stubblefield, Had Wireless Phone in 1902 — Predicted Broadcasting

Special to the *New York Times*.
Murray, Ky., April 23 —
Death, which came several days before his body was found, has ended the dreams of Nathan Stubblefield, who in 1902 made great strides in what he called "wireless telephony" and which has become radio.
Mr. Stubblefield, who was 65, lived like a hermit in a small house in Calloway County, near here. His wife and five children long ago left this section and where they are is not known. He had refused efforts to help him, even declining aid from his brother, Walter.
Mr. Stubblefield in an interview given in 1902 forecast radio and its ramifications, including broadcasting, and declared he had solved the problem of sending the human voice through the earth, which he said he had started on ten years previous, before he had heard of Marconi's efforts.
Twenty-six years ago he took his apparatus to Philadelphia and there gave a demonstration which is said to have been successful as far as it went, but could not interest sufficient capital to market or further his plans. He became a disappointed and disillusioned inventor.

The Legend Begins

If you stabbed L.J. Hortin in the chest, his aorta would have pumped out printer's ink because he was a newspaperman to the heart. He was just the sort of person that Rainey Wells wanted to run the fledgling journalism program at Murray State Teachers College. Wells, who witnessed early Stubblefield wireless tests, was instrumental in bringing the new Normal School to Murray and had become its second president. Kyle Whitehead started the program but left after one year. Wells convinced Hortin to take a vacation from his job at the *St. Louis Post Dispatch* and earn some extra money teaching summer school at Murray. By the end of the term, Hortin resigned from his job and took up a full time teaching position beginning in the fall of 1928.

Hortin taught basic writing, editing, and management courses and served as editor of the twice-monthly *College News* and adviser to the college yearbook. A demanding but fair instructor, he saw no sense in teaching students how to write then showing them what to write about. He preferred to set up a news laboratory where they started with leads that they could investigate, and learned to gather facts, write and edit along the way. In this manner, the students mastered basic techniques and ended up with finished stories ready for publication.

To set up the news laboratory, Hortin needed material for assignments. He naturally chose local news because his students had convenient access to the people, places, events, and background information that served as resources. Hortin hit upon the Stubblefield story early in his search. Nathan's pathetic death and the tales of his life were still fresh memories in the small town. Perhaps Dr. Wells gave Hortin a tip. He got to know Nathan's cousin Vernon Stubblefield who showed him photographs and memorabilia. Hortin's curiosity soon grew into fascination. The newspaperman's blood began pumping as he contemplated the news values and human-interest angles. It didn't hurt that his former employer, the *St. Louis Post Dispatch*, had started the story in 1902 and was a potential resource.

Noted historian Forrest Pogue was a member of Hortin's first group of journalists who dug up the Stubblefield Story. He recalled that Hortin's rules were simple: "Investigate. Find the facts. Verify the facts. Draw your own conclusions. Then you can start writing." The small band of young newshounds eagerly followed the trails and slowly their accounts began to emerge in the *College News*. In the fall of 1929, Hortin added a course in feature writing with the plan to turn some of the better Stubblefield pieces into syndicated stories for the local press and news wires. If the plan worked, the students would gain a sense of accomplishment and earn some pocket change at the same time.

Meanwhile, Vernon Stubblefield convinced Hortin that, based on the information at hand, Nathan Stubblefield invented radio, and he found that the *New York Times*, the *World Almanac*, and other publications agreed. In the 1892 demonstrations to Rainey Wells, Will Mason, and others, Nathan had clearly transmitted voice by wireless telephone before anyone else. He was also ahead of Marconi, who never attempted anything but telegraph messages, by at least three years. In that era, it was all called wireless. The term radio came later. To Hortin, the difference between what Nathan did and what Marconi and others achieved was semantic, not technological. As Hortin explained in his 1951 article for *Broadcasting* magazine:

> ... *telegraphy* is different from *telephony*. Hence wireless telegraphy and wireless telephony are different inventions. Telephony has to do with the transmission of *sound*, while telegraphy does not. The radio of today is understood primarily to refer to the transmission and reception of sound.

So Hortin began a crusade for Stubblefield's recognition that would last to the end of his life. The first task was to get the town to commemorate its inventive genius. Hortin proposed a monument on the site of Nathan's home and workshop where those 1892 tests took place. Working through the Chamber of Commerce, and later the Rotary Club, he floated the idea of buying part of the old farm, now adjacent to the college, and turning it into a park and museum. The onset of the Great Depression thwarted his fundraising plans.

Collecting a few dollars here and there from Nathan's relatives and other interested parties and finally paying most of the bill himself, Hortin raised enough money for a small marble monument to be located on the college campus, across the road from Nathan's original home place. The inscription read:

> Here in 1902 Nathan B. Stubblefield — 1860 - 1928, inventor of radio — broadcast and received the human voice by wireless. He made experiments 10 years earlier. His home was 100 feet west.

Stubblefield Monument

Original design, 1929

Pogue Library

Pogue Library

Dr. Rainey T. Wells, left, and L.J. Hortin at the dedication of the Stubblefield Monument, Murray, March, 1930

Illustrations of the 1908 desk telephone and a receiver from the 1902 Philadelphia demonstration decorated the marble face. Hortin scheduled a dedication ceremony for March 28, 1930, the second anniversary of Nathan's death.

To prepare for and promote the event, Hortin produced two feature articles for the trade press, based on the facts that he and his students had uncovered. One appeared in the *St. Louis Post Dispatch* on the cover of its February 23, 1930, Sunday magazine. The lurid headlines read: "TRAGIC END of the MAN NOW HAILED as the INVENTOR of RADIO" and "Nathan Stubblefield Broadcast Human Speech by Wireless in 1902 — And Died Two Years Ago, Half-Starved, Penniless, and Unattended." The article itself was largely the reprint of the January 12, 1902, fcature in the *Post Dispatch*.

At about the same time, the other article appeared, inaccurately attributed to L.T. Horton, in *Kentucky Progress Magazine*, with the title "Murray, Kentucky,

Kentucky Progress Magazine

Murray, Kentucky, Birthplace of Radio

By L. T. HORTON

"BIRTHPLACE of Radio" is the honor that is being claimed by the citizens of Murray, Kentucky, who will dedicate a marker on the campus of Murray State Teachers College on March 28 in honor of Nathan Stubblefield, the first man to broadcast and receive the human voice without wires.

After proving to the world in 1902 that he could successfully transmit the voice by wireless, this Murray genius came home to die in a little hut near the city two years ago.

What price glory? Although he undoubtedly gave the world its greatest invention, the radio, he failed to get the honor due him. He wanted glory, for he wrote to his cousin Vernon Stubblefield: "You and I will yet add luster to the Stubblefield name. N. B. S."

Now the little tobacco and college town of Murray, Kentucky, is trying to add that luster to his name by erecting in his memory a marker near the ruins of his old home. The tardy memorial will be dedicated March 28, 1930, exactly two years after his death.

Nathan B. Stubblefield and his son "at the microphone" for the Philadelphia demonstration, May 30, 1902.

"Hello, Rainey"—these were the first words conveyed by ether, the first radio message. To Dr. Rainey T. Wells, then a young attorney, Nathan Stubblefield in 1902 broadcast without wires the above message across a swampy Kentucky wood, now a beautiful campus. His only equipment consisted of a "crazy" box, some telephone equipment, two rods, and coils of wire.

When news of the "crazy" Kentuckian's invention reached a Marconi-dazed world, "Eastern scientists and capitalists wearing diamonds as large as your thumb" flocked to the farm home of Nathan Stubblefield.

"I saw a written offer of $40,000.00 for a part interest in the invention," stated Dr. W. H. Mason, Murray surgeon, last week. Half a million dollars was refused by Stubblefield because he thought his invention worth twice that sum.

On January 1, 1902, he demonstrated before a thousand people that the human voice could be broadcast and received without wires. A St. Louis Post-Dispatch reporter, in a full-page article of January 12, 1902, says:

"However undeveloped his system may be, Nathan B. Stubblefield, the farmer inventor of Kentucky, has assuredly discovered the principle of telephoning without wires. Today he gave the Sunday Post-Dispatch a practical demonstration of his ability to do this and discussed his discovery as frankly as his own interest and self protection would permit."

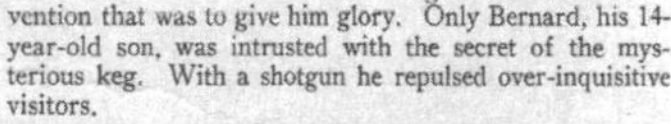

A soda keg first housed the invention that was to give him glory. Only Bernard, his 14-year-old son, was intrusted with the secret of the mysterious keg. With a shotgun he repulsed over-inquisitive visitors.

Broadcasting and receiving radio messages on the Potomac River, March 20, 1902. (The first marine demonstration of wireless telephony.) Stubblefield is third from the left end of the steamer "Bartholdi." Note the two aerials.

Pogue Library

Article by L.J. Hortin, 1930
The first comprehensive summary of Stubblefield's life story, where Hortin coined the phrase "Birthplace of Radio."

Birthplace of Radio." This piece is significant for two reasons. It was the first attempt to summarize and explain Nathan's life story, so it has become the most quoted, and misquoted, source for subsequent writers. Hortin himself continued to revise, republish, and excerpt parts from this article for another 60 years. The entire text is included as Appendix D. But more importantly, the phrase "Birthplace of Radio" caught on in Murray. Over the next four decades, the community internalized the concept and attempted to exploit it for progress and economic gain.

Hortin preferred to stick to the facts. Although he was an entertaining writer, he was careful to document and attribute sources for the more outlandish statements:

> When news of the "crazy" Kentuckian's invention reached a Marconi-dazed world, "Eastern scientists and capitalists wearing diamonds as large as your thumb" flocked to the farm of Nathan Stubblefield. "I saw a written offer of $40,000 for a part interest in the invention," stated Dr. W.H. Mason, Murray surgeon, last week.

Acknowledging that Stubblefield folklore was already present, Hortin used precise language to categorize the intriguing, yet unverified, parts of the story:

> Strange tales are told by the superstitious Kentuckians about his last days. Living alone in his little hut, six miles north of Murray, near Almo, he was working on wireless plans. Wireless lights appeared up in trees, on the end of steel rods, and along a woven wire fence, according to rumors spread by the people of Almo, Kentucky. Voices, coming apparently from the air, were frequently heard if trespassers are to be believed.

Hortin later admitted that he had never met anyone who had actually seen the lights.

The dedication ceremony for the Stubblefield monument came off as scheduled. Rainey Wells, Will Mason, and others told of what they had witnessed. Two of Nathan's daughters, Victoria and Pattie, were present.

Pogue Library

L.J. Hortin with his Murray State journalism students, 1932, who first investigated and reported the Stubblefield story Historian Forrest Pogue is third from right.

Pogue Library

The Murray State *Shield*, 1932
Cover design with microphone and
broadcasting motif

Out of appreciation for their mentor, the student editors of the 1932 Murray State yearbook, *The Shield*, dedicated the volume to Nathan. Its cover embossed with an art deco microphone projecting thunderbolts into the air, the inscription read:

Dedication

The Shield staff of Murray State Teachers College respectfully and reverently dedicates this yearbook to the immortal memory of Murray's genius, Nathan B. Stubblefield, inventor of the radio. Years before the dream of a college here became a reality, this citizen of Calloway County toiled through long hours of sleepless nights to produce the greatest invention of all time—the radio. Nathan B. Stubblefield, scholar, farmer, genius, eccentric, experimented with his "wireless telephone" long before Marconi, DeForest, and others had become known for their contribution to wireless broadcasting and reception. Stubblefield's name has become associated with this college campus and its traditions. His home, the scene of his experiments, was adjacent to the estate on which this magnificent college is situated. Over these very grounds this Kentucky immortal first flashed to an unbelieving world the news that there was "music in the air." Though he lived in poverty, loneliness, and disappointment, Stubblefield shall live forever in the hearts of the students, faculty and alumni of Murray State Teachers College. Even as his rude radio carried messages of hope, music, and culture, so shall this book communicate to a receptive world the aspirations and achievements of this college.

Following the dedication, the yearbook contained two pages of pictures of Nathan, Bernard, the inventions, and L.J. Hortin with his student journalists.

At this point, Hortin shifted gears and became more of a promoter than a journalist. He and Vernon Stubblefield launched a two-pronged attack to get national recognition for Nathan's achievements. Hortin's dream remained the establishment of a park and larger memorial at the old home place. He petitioned the National Park Service in 1934 to no avail. A bill before the Kentucky General Assembly to appropriate $16,000 for the Stubblefield Memorial failed in 1936. The next year, Bailey Wooten, Director of Kentucky Parks, wrote to Hortin suggesting that city of Murray buy the property and donate it to the state. Working through the Chamber of Commerce and the Rotary Club, Hortin was unable to raise enough money for the scheme. It is unclear whether economic circumstances or lack of interest was the primary obstacle.

On the other hand, the publicity campaign was going well, and had migrated from print to the airwaves. Vernon convinced the program director at WSM, Nashville, to do a feature on Nathan. A clear channel AM station at 650 kilohertz, WSM had a night time signal that covered most of the eastern United States and parts of Canada. It was also one of the more important NBC affiliates. WSM aired the program on December 11, 1936. L. J. Hortin told most of the story himself. The broadcast ended:

> As the telegraph and the telephone are separate and distinct and distinct inventions, so are the wireless telegraph and wireless telephone different inventions and mechanisms. The "wireless telephone" is the "radiotelephone,' or, simply, the radio. We submit that Nathan B. Stubblefield was the first to discover, invent, manufacture and demonstrate equipment for broadcasting and receiving the human voice and music by wireless. He invented the radio.

Unbeknownst to HORTIN and Vernon, they were now very close to raising the funds for the Stubblefield Memorial. The potential source was one of America's major corporations.

In the 1930s, RCA was as dominant in the radio industry as Microsoft is in today's computer business, and David Sarnoff was the man who ruthlessly wielded that power. The RCA president was fond of saying: "I don't get ulcers. I give 'em." For a time in the middle of that decade, Sarnoff seriously considered underwriting a Stubblefield memorial in Murray, commemorating the inventor's wireless telephony experiments.

The suggestion first arose in 1935 when RCA, owner of the NBC Radio Network, became aware of the plan for a Stubblefield historical site. In November, an ad for the Citizens Union National Bank in the *Louisville Courier*

Journal caught David Sarnoff's attention. Entitled "Radio —- A Kentuckian's Gift to Mankind," it read in part:

> In 1892, across a swampy Southern Kentucky Woodland, now the campus of Murray State Teachers College, was sent the world's first "raidio" message. "Hello Rainey" were the words transmitted by Nathan B. Stubblefield, the inventor to a young Dr. R.T. Wells, attorney and friend. The only equipment for the first broadcast was a mysterious box, some telephone apparatus, two rods, and coils of wire, but the genius of an obscure Kentuckian had performed the greatest scientific "miracle" of all time.

As an aside, the origin of the description of Nathan's farm and nearby woods as "swampy" is obscure but inaccurate. There are no wetlands in the vicinity now, and the photographs of the house and orchard show a well-drained landscape. The St. Louis reporter who traversed the area in 1902 makes no mention of wading through any bogs. Yet this description persists in Stubblefield lore.

Pogue Library

1935 Newspaper ad that caught David Sarnoff's attention

Intrigued by the story, Sarnoff referred the matter of RCA's participation in a Stubblefield memorial to George Howard Clark. Clark was in charge of RCA's Show Division, a traveling exhibit devoted to the history of radio and RCA's contributions to the industry. An MIT graduate, Clark had started a personal collection of wireless and radio memorabilia in 1902. He worked for Stone Telegraph and Telephone, then became a civilian wireless operator for the US Navy in 1908. After World War I, he went to work for Marconi and joined RCA a few years later.

Clark already had an extensive file on Stubblefield. Based on his understanding of the history of radio technology, he responded negatively to the idea. In a detailed report, Clark argued that "Nathan B. Stubblefield was an eccentric, uneducated dabbler in electricity." Moreover, his inventions were based on either ground conduction or induction. Clark said:

> One thing is sure, and that is, Stubblefield did not have any radio frequency features in his work. A lawyer who knew him well talked with me while I was at the World's Fair in 1933, and told me much about the Stubblefield claims, and when I told him that it was clear that the old man was using an induction scheme, he admitted that this was the fact.

Instead, Clark suggested a pamphlet or feature "on Communication without Wires before the use of Hertz Waves, [in which] Stubblefield's work [could] be fairly described, and due credit given to the eccentric old recluse for at least having worked on the problem." Sarnoff lost interest.

The idea resurfaced in early 1937, after the radio program on Stubblefield aired over powerful affiliate WSM and an article about it appeared in *Broadcasting* magazine. Once again, RCA considered building a national shrine in Murray. NBC executive Frank Mason suggested: "I don't know whether Mr. Sarnoff recognizes Stubblefield as a factor in radio, but if he does there may be the germ of a good promotion idea here." Clark responded:

> If sufficient precaution could be taken to have the public understand that the work of Stubblefield in no way ante-dated that of Marconi; that the two

> were working along totally different lines; the former by a scheme which science later showed to be impractical, ... there might be good publicity in making his burial plot a national radio shrine.

Clark went on to state that "the trouble lies in the nomenclature." All wireless telephones and telegraphs of the late 19th century were not radio. Although in Great Britain the term wireless persisted, the US Navy adopted the term *radio* in 1912 to differentiate wireless by electromagnetic waves from any technologies prior to Marconi. Since RCA held the US rights to all Marconi patents and they were still tied up in legal actions over infringements, it was in the company's best interest to maintain the Italian inventor's priority. Sarnoff agreed with Clark again.

During this period, there is an interesting series of letters between Hortin and Reed Hilty, a studio page at NBC. Hilty apparently heard of Stubblefield from the internal grapevine and contacted Hortin in 1937 with the idea that the studio pages might create and produce a radio drama about Nathan and broadcast it on the network. Hortin was enthusiastic and continued the correspondence until June 1938 when a program director at NBC turned the idea down. Hilty commiserated with HORTIN, claiming that the decision "was not altogether fair" and that "politics plays a large part in radio as well as all other fields." He went on to blame Marconi's son Julio for the NBC action.

Hortin and George Clark also had a brief correspondence in 1942. Clark wrote to the college requesting that they exchange copies of documents and photographs relating to Stubblefield. HORTIN replied favorably to the request with no indication that he knew of Clark's role in the internal discussions at RCA regarding the Stubblefield Memorial.

Meanwhile, Vernon Stubblefield was able to interest other radio networks in programs about his cousin Nathan. He and Nathan's oldest son Bernard, who now lived in New York City, appeared on a national broadcast of the series "We the People" on CBS, July 30, 1940. While in New York arranging for the broadcast, Vernon also sent syndicated columnist Walter Winchell a telegram: "Would it interest you to know the first man to broadcast human voice and music has been found in New York City. Living and documentary evidence to sustain claim." Winchell didn't respond. But Vernon succeeded in placing a program about Nathan on the Mutual Radio Network series "The Answer Man," which originated at WJR, Detroit on August 21, 1940.

Edward Freeman, one of Hortin's former students, took up Nathan's cause with a series of colorful articles written around 1939. Freeman was not very careful with his facts, often came to erroneous conclusions, and relied heavily on local Stubblefield lore. In one piece he wrote:

> At 22, Stubblefield demonstrated to a Murray audience that earth currents of electricity would excite a compass needle, although several yards lay between the compass and the electromagnetic generator. …
>
> He worked out some theories, but in the meantime he saw the need for telephones in Murray. Bell's patent on the telephone had not expired, so Nathan developed a model of his own. It was rather a simple device, and the amazing thing about it — the significant thing! — was that the phones were not connected by a wire. Bell's were, but not Stubblefield's. He used a small waxed cloth cord. Was this perhaps the earliest form of radio ever to come into the world?
>
> The telephone was about 12 inches square, and it stood about four inches from the wall. Stubblefield had some kind of batteries he'd developed inside. From the box extended the transmitter, an oyster can with a stiff waxed cloth over the open end for a drum. A small wooden hammer was handy to peck on the drum, and the sound was repeated at the other end of the line.
>
> "One night in 1885," [Duncan] Holt said, in telling how the telephone worked, "a bunch of us were gathered at the Dale and Martin's Drug Store on the corner, and somebody wanted to get Dr. Hart

> for a sick child. Those were the only two telephones in town — at Dale and Martin's and at Hart's. Well someone phoned up there — some three hundred yards, and mind there was only a waxed string connecting the phones — and we all heard Dr. Hart say the kids were making such a fuss that he couldn't understand what we were saying."

In another article, Freeman described the 1902 demonstration on the Murray courthouse square but dated the event 1892. These errors crop up often in subsequent stories by other writers. Because Freeman went on to important editorial posts at the *Louisville Courier Journal* and *Nashville Tennessean*, anything he wrote had credibility and authority, even if it was wrong.

In 1940, Stubblefield's son Nathan Franklin and Conn Linn, the attorney who invested in the 1908 patent, briefly considered legal action to prove priority over various patents in the lucrative radio business. Their New York patent lawyers discouraged this suit. Nathan's patent had expired. In the event that they could prove anticipation, they would win no monetary damages. They decided not to pursue the case. Over time, this incident got twisted into a tale that the New York Supreme Court passed down a judgment in Nathan's favor, but that body never rendered such a decision.

Interest in Stubblefield waned for a few years, perhaps because people in Murray and the nation had focussed their attention on World War II. Hortin and Vernon did make some progress in 1944 when they convinced the Kentucky General Assembly to pass a resolution proclaiming that Nathan was "the father of radio" and a man of "outstanding scientific achievement and public service." But there was still no appropriation forthcoming for the Stubblefield Memorial.

In 1947 Vernon Stubblefield provided background information for a radio drama on Cincinnati's powerful WLW AM. Previous print and broadcast versions of Nathan's story had been basically documentaries, so this fictional rendition marked a departure from the norm and the start of a new direction for Stubblefield folklore.

The program about Nathan aired Thursday evening, January 15, 1948, as part of the series *Builders of Destiny*. In a letter to Vernon Stubblefield, B.M. Matteson, continuity editor at WLW, remarked: "You will note certain dramatic liberties that have been taken in the development of this script. But we have retained the facts as we found them in the material you sent to us." Scriptwriter Dave Brown also laced the play with broad rural humor that comes across as poorly crafted hillbilly stereotypes. One conversation between two anonymous citizens of Murray reads:

> VOICE 1: I say it's against all the laws of God and man ... You just can't talk without wires.
>
> VOICE 2: Yeah? Seems to me like you said once you couldn't talk WITH wires ... but we're doin' it, aren't we?
>
> VOICE 1: That's different. You gotta have some way of conductin' the electricity.
>
> VOICE 2: But Stubblefield uses the electricity in the ground instead of electricity in a wire ... and whatever he's got in that box.
>
> VOICE 1: Uh-huh. And what do you suppose he's got in that box, anyway?
>
> VOICE 2: I don't know .. That's th' secret.
>
> VOICE 1: Yeah. And you know why ... cause there ain't NOTHIN' in that box ... It's a fake.
>
> VOICE 2: What! Nothin' in it. You ... really think so?
>
> VOICE 1: Course I think so. Has anybody ever seen IN that black box?
>
> VOICE 2: No. Don't rightly guess anybody has.
>
> VOICE 1: Well, there you are ... That proves it. He won't let nobody see IN the box 'cause there ain't nothin' IN that box.
>
> VOICE 2: You know, you may be right at that.
>
> VOICE 1: Course I'm right. And if you ask me that box is a crazy box. Jest a plain empty box to fool CRAZY people that think folks can talk all around the country without wires. Yes sir ... th' whole thing's crazy.

The "dramatic liberties" are apparent in the next act. Compounding Freeman's error of dates, Brown stages a scene between Nathan and a mature Rainey Wells in the fall of 1892 in the latter's nonexistent law office. Then the drama proceeds to a demonstration of equipment that Nathan would not invent for at least eight years. To top it off, the description of this event totally ignores the facts as Wells told the story. Consider this dialogue between Wells (who was 17 in 1892) and Bernard Stubblefield (who was 5). Nathan has sent them out to the orchard with a telephone receiver:

> BERNARD: All right, Mr. Wells, this ought to be a good place don't you think?
> WELLS: Suits me all right young man. What shall I do? Can I help hook up anything?
> BERNARD: No sir. I'll just stick these rods in the ground. This one can go right here. [Sound Effects, Grunts from Bernard] There. That ought to be deep enough. And the other one can go over here.
> WELLS: Is that all you have to do?
> BERNARD: Yes sir. Except to hook these coiled copper wires up to the transmitter boxes. But that won't take a minute sir.
> WELLS: I see.
> BERNARD: See, I just attach the wires here to these binding posts and then we're ready to go
> WELLS: How do you happen to know all about this, sonny?
> BERNARD: Oh, dad's explained it all to me. Well, there it is ... Everything's ready to work now.
> WELLS: What do I do?
> BERNARD: Here, you hold the telephone, while I call to dad.
> WELLS: All right
> BERNARD: (CALLING) Hello. Hello, dad. All ready at this end. All ready. Are you listening, Mr. Wells?
> WELLS: Uh-huh. Only I don't hear anyth....
> NATHAN: (FILTER MIKE) Hello, Rainey. Do

you hear me

WELLS: (SURPRISED) Hey, what was that? Where'd that come from?

BERNARD: (LAUGHING) That was Dad, Mr. Wells. He's talking to you over the wireless telephone.

WELLS: But ... but it can't be. It's ... it's a hoax. You're playing a trick on me.

The script also includes the ship-to-shore conversation in Washington, 1902, with Nathan onboard talking to Bernard, who did not make the trip, on shore. Unlike the actual test where Nathan had to rig up more batteries hurriedly for more transmission power, this one goes off without a hitch. The play avoids any mention of the Wireless Telephone Company of America disaster. Instead, the narrator skips ahead to the 1908 patent and concludes the drama with rhetorical splendor:

> Offers of great sums of money ... some say as much as half a million dollars were made for the rights to the invention, but Mr. Stubblefield turned them all down. Why? Men still living, who were close to Nathan Stubblefield, are unable to explain why he failed to market his invention. All that is known is that he died, in poverty and sorrow, in March, 1928. When found in his tiny, unheated shack, he had been dead for several days ... and so passed the man who doubtless helped guide other inventors to the devices that made radio a reality and this program possible. And so tonight, we honor a man who in his lifetime was denied the honor due him for a great invention ... and add the name of Nathan Stubblefield of Murray, Kentucky ... to the roll of ... BUILDERS OF DESTINY.

In the summer of 1947, L.J. Hortin moved to Athens, Ohio. Recognizing his accomplishments as a journalism educator, Ohio University hired him away from Murray State. He eventually became Director of the E.W. Scripps School of Journalism and started respected Master's and

Ph.D. programs at Ohio.

Hortin's legacy as an educator at Murray State was impressive, but he was unable to control the direction that Nathan's tale was taking. Although he originally tried to tell a good story based on verifiable facts, his incursions into sentimentality and sensationalism undeniably turned the story in to folklore. Fueled by misinformation, factual errors, and tall tales, Stubblefield mythology soon took on a life of its own. Hortin's last comprehensive attempt to tell Nathan's story his way was the article "Did He Invent Radio?" in the March 19, 1951, issue of *Broadcasting*. Although he continued to publish annual commemorative articles about Stubblefield in the local press and returned to Murray State in 1967, Hortin was never again as active in building and promoting the Stubblefield legend as he was in the 1930s and 1940s. Others took up that task and turned it in new directions.

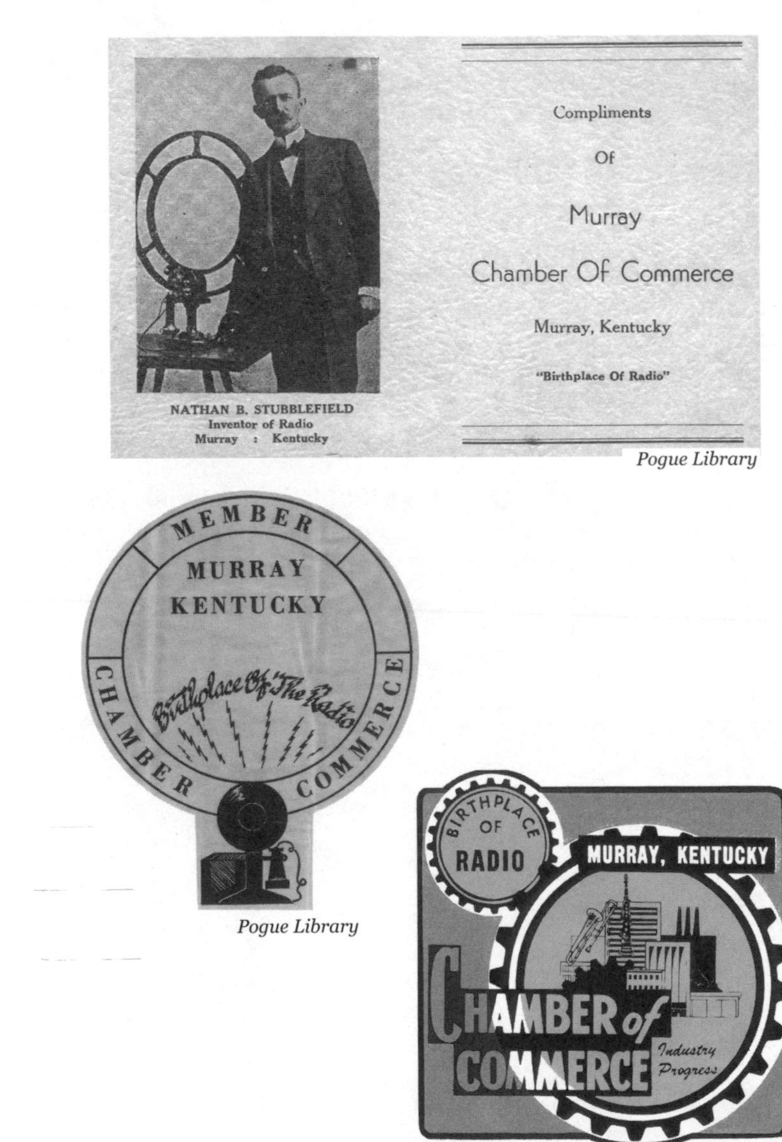

Pogue Library

Pogue Library

Pogue Library

Murray Chamber of Commerce promotional materials through the years, evolving from biography to celebrating technology to making connections with a high-tech future.

The Birthplace of Radio

On September 12, 1946, page one of the *Murray Ledger & Times* carried this story, with the bold headline:

> BROADCAST PERMIT SOUGHT FOR 'BIRTH-PLACE OF RADIO' — GEORGE E. OVERBEY HEADS MURRAY COMPANY
>
> Officials of the Murray Broadcasting Company have announced that application has been filed in Washington for permission to operate a broadcasting station here. George E. Overbey is president of the company; W.G. Swann vice president and M.O. Wrather secretary-treasurer. Assisting the company in its application is Neville Miller, former mayor of Louisville. Hearing on the application, submitted September 9, will be held September 26, 27, and 28.
>
> Federal approval of the application will be the culmination of years of effort by various local organizations to bring a modern radio station to Murray and to obtain suitable recognition of Murray as the "birthplace of radio."
>
> To Honor Stubblefield
> Although the call letters of the station have not been determined, members of the corporation have expressed their desire that they might be WNBS, the last three letters being the initials of Nathan B. Stubblefield.
>
> Stubblefield, Murray, who died in 1928 is widely credited with having invented and demonstrated the first broadcasting device at the County Court Square in 1902. Several years ago, the story of Stubblefield's early efforts were [sic] broadcast on the nation-wide program "We the People." The story was related by his son, Bernard Stubblefield, and by Vernon Stubblefield, Murray druggist and relative of the pioneer inventor. Numerous authoritative publications give recog-

> nition to Stubblefield for his leadership in the field of radio.

This was an excellent business opportunity for Overbey, an attorney and soon-to-be state senator, and his partners. After World War II, the FCC tackled many regulatory issues postponed by the priorities of military communication. Among these matters were the establishment of television, restructuring FM radio, and the expansion of local service for AM radio. In the latter case, the commission actively allocated frequencies, sought applications for licenses, and sent staff around the country for hearings on competing applications for the same channel. In more than 25 years up to January 1, 1946, the FCC authorized 1004 AM stations. In the next five years, it would authorize 1228 more.

Most of the new radio stations were in small towns like Murray, Kentucky, and Paris, Tennessee, the two communities seeking the service at 1340 kHz. Because AM signals travel farther at night, the majority of these local channels were low power and daytime-only so that their signals would not interfere with existing stations, primarily in cities. 1340, however, was a fulltime channel and more valuable because of its extended hours, especially in the winter when radio audiences were larger. The FCC had already allocated a daytime channel to Paris, and a group of local businessmen had already secured a construction permit with plans to be on the air in mid-1947. This prior allocation probably weighed against the applications of two groups from the Tennessee town and in Murray's favor. Under a regulatory principle called *localism*, in comparative hearings the FCC tended to choose communities without radio service and owners with no other radio properties. Nevertheless, the Murray Broadcasting Company (MBC) had to prove that it was financially and technically capable of operating a radio station for the initial seven-year license term.

Although MBC advanced the Stubblefield connection as part of its argument, it was clear that the primary objective was to operate a profitable radio business. In a follow-up story January 9, 1947 about the hearings, the *Murray Ledger & Times* wrote:

> George E. Overbey, president of MBC, has stated that his organization has every confidence that the FCC hearing will provide ample proof that the need for a suitable radio station in Murray will justify the construction of the station here. However should Murray fail in its petition, Paris might find itself with three stations, and Murray, which gave radio to the world, still unable to provide Calloway industry and other interests with a badly needed broadcasting concern. ... Overbey has requested that any local musicians, bands or entertainers interested in radio work contact him immediately.

Finally, after the FCC "changed its mind more times than a bargain basement shopper" about the hearing dates, the commission staff arrived in Murray on January 23, 1947 to hear the case for allocating 1340 kHz to the town and awarding MBC the construction permit. The hearings lasted two days, and most of the testimony dealt with the technical specifications of the transmitter and tower, the financial arrangements, plans for programming, operating policies, and the benefits to the community. These were the criteria that the FCC would judge. The mayor, the county judge, the president of Murray State, and other community leaders appeared. But the star witness was Dr. Rainey Wells. His testimony was front- page news in the *Murray Ledger & Times* of January 30, 1947:

> DR. WELLS TELLS OF BIRTH OF RADIO
> AT FCC HEARING ON MURRAY STATION
>
> As the long-awaited Federal Communications Commission hearing on Murray Broadcasting Company's petition for boradcasting [sic] rights was held here Thursday and Friday last week, Murrayians had the privilege of hearing one of Calloway's most distinguished residents tell the story of what all available records indicate was

> the first broadcast of the human voice in the history of the world.
>
> Dr. Rainey T. Wells, former Calloway County attorney and president of Murray State College, told the FCC representative how Nathan Stubblefield gave a demonstration of his broadcasting device in 1902 and added, "As far as I know. That was the first conversation carried by what we now call radio. ... I was Stubblefield's attorney in 1901 for the purpose of securing patents for his apparatus," he said. The onetime general attorney for the Woodmen of the World insurance society said that he was called to Stubblefield's home in the summer of 1902 to participate in the demonstration. Stubblefield entered what "we would call a broadcasting booth today" and spoke to Dr. Wells from distance of 2000 to 3000 feet. Dr. Wells said that he was highly skeptical of the invention at that time and moved about with the receiver during the experiment to eliminate the possibility of concealed wires or other trickery. "The apparatus worked to perfection," Dr. Wells declared.

There are some interesting points about this statement. Wells did not mention the earlier 1892 demonstration that he had witnessed as a 17 year-old boy. That experience was the subject of his newspaper and magazine interviews in 1935. Instead, he chose a similar event from the summer of 1902, after Nathan's split with the Wireless Telephone Company of America. Perhaps he felt that his status at the later date, as an attorney representing a client, would produce an affidavit that carried more weight with FCC staff lawyers.

The hearings had an almost immediate impact on the local media. That summer, in anticipation of new competition, the *Murray Ledger & Times* expanded from weekly to daily publication.

It is unclear how important Wells' statement and the Stubblefield connection were to the ultimate decision to allocate the channel to Murray and grant the construction permit to MBC. Construction began in February 1948, and the transmitter was ready for full-scale tests by late June. On July 1, Overbey got a telegram from the FCC authorizing the station to commence broadcasting. At 3 minutes past 7 that evening, Calloway County first heard the now-familiar station ID: "This is WNBS, Murray, Kentucky, the Birthplace of Radio." L.J. Hortin, back in Murray for the summer, was on hand and was introduced as "one of the tireless workers who had been instrumental in having Murray recognized as the birthplace of radio."

Although Nathan's initials were the call letters, no picture or any reference to him was to be found at WNBS. Meanwhile, Nathan's unmarked grave had grown up in weeds so badly that it was impossible for most people to locate. The United Daughters of the Confederacy chapter in Murray took up the cause improving Nathan's burial site. In February 1947, it asked the Chamber of Commerce to clean up the Bowman cemetery and contributed $75 to start a fund to erect a marker on Nathan's grave. Apparently, the Chamber took no action, and more than a year later, the Boys Scouts volunteered to clear away the weeds and underbrush. In May 1948, the U.D. of C. provided a progress report, reported in the *Murray Ledger & Times*:

> The grave of radio's inventor, Nathan B. Stubblefield, will no longer be marked by only an iron stake, it was announced by Mrs. W.W. McElrath, president of the local chapter of U.D. of C. The group made plans to erect a fitting monument at the grave of Stubblefield in the Bowman Cemetery, approximately one mile north of Murray. "It will be a modest monument," said Mrs. McElrath, "because Mr. Stubblefield was a modest man and we believe that is what he would want." ... Present plans include placing a marker on the highway showing the location of Stubblefield's grave. The entrance to the cemetery will also be improved and the grounds beautified. The grave of the inventor of radio is expected to be one of the leading tourist attractions in this area.

Nathan finally got his tombstone in 1952, with his children footing most of the bill. For many years, volunteers cleaned out the cemetery.

Ostensibly, the Murray Chamber of Commerce was more interested in Nathan as part of a marketing plan to encourage civic pride, recruit new members, and attract industry. The Chamber distributed a window decal to members that celebrated the "Birthplace of Radio." It created a brochure, with text written by L.J. Hortin, that summarized the Stubblefield saga and distributed it freely. L.D. Miller, Executive Secretary of the Chamber explained this enterprise in a letter to radio producer William Wells:

> We are confident that Murray has received more publicity from Nathan B. Stubblefield and his great invention than from any other thing that has happened in Calloway County. Through efforts of this kind we can continue to publicize Murray and Nathan B. Stubblefield as one of the greatest inventors of all times.

A.G. Weems, a reporter from the Memphis *Commercial Appeal*, visited Murray in May 1948 and filed a story about the community and its local hero:

> The first thing they tell you when you hit this town is that Murray is the birthplace of radio, and they'll have a lot of facts and figures and quotes from reliable sources to prove it, Mr. Marconi to the contrary and notwithstanding. But that's not all they'll tell you. I talked with several "I'm-Proud-To-Live-Here" residents, Chamber of Commerce officials and Mayor George Hart, and in two hours they stuffed me with enough information about this town to write a book.

After he ran down the basic Stubblefield facts from the Chamber brochure, Weems devoted the bulk of the article to boosterism. The topics included the new Tappan Stove Company plant, the surplus of labor, the availability of electric power, the modern water and sewer system, the local hospital, and the "great and beautiful Kentucky Lake." In closing, Weems stated with enthusiasm:

> "The Murray of today is the fastest growing town in Kentucky," says the Murray Viewbook published last year. And, for one, I'm convinced that statement is no exaggeration.

This was just the sort of publicity that the Chamber sought with its "Birthplace of Radio" campaign. The local business community was willing to contribute to anything related to Stubblefield so long as it brought attention to Murray and Calloway County, but otherwise had little interest in remembering Nathan. In 1952, L.J. Hortin again tried in vain to form a non-profit Stubblefield Foundation to fund his dream of a suitable memorial. On the other hand, when a minor Hollywood luminary offered to take up Nathan's cause, the town rolled out the red carpet.

In 1956, L.D. Miller contacted actor Marvin Miller (no relation), told him Nathan's tale, and convinced him to come to Murray to learn more. At the time, Marvin was working on the popular TV show *The Millionaire*. The plot of this weekly half-hour drama revolved around a wealthy but eccentric man who chose people at random and gave them checks for a million dollars. Each episode followed the new millionaire to see how the money changed his or her life. Marvin played Michael Anthony, the errand boy who delivered the check. Like many early TV performers, he had a former career in radio and good contacts within that profession on the West Coast.

Marvin Miller came to Murray in 1957 and was feted royally by the Rotary Club, the Chamber of Commerce, and other groups. Although he had no way to work Nathan into his television series, he promised to organize a committee of radiomen in Los Angeles and elsewhere to commemorate the Stubblefield legacy. The immediate result was another radio feature, this time on the *Behind the Story* series on KHJ, Los Angeles. Beyond that, Marvin Miller was unable to stir up any interest. In 1961, he wrote to

James Johnson, who had become Executive Secretary of the Chamber, with an interesting but unsubstantiated conspiracy theory:

> For over a year I did everything I could to stir up interest in TV and radio circles, particularly in reference to the annual "Stubblefield Award" for the development which benefited some branch of the radio or TV industry the most during the preceding year. At first, all the publicity experts who heard about it were wildly enthusiastic. Then, interest suddenly waned! For a long time, I was unable to find the reason — until finally my own publicist discovered it. The networks, the motion picture companies and broadcasting stations, and secondarily, even the advertising agencies and production companies, are all afraid to do much talking about Stubblefield — lest it should appear that we have unloosed a hornet's nest, and all the current physical patents in the industry should be declared null and void in court! I did my utmost to assure everyone that there was no possible chance of late claims being lodged against the present patent-holders — that N.B.S. had cooked his own goose when he failed to patent his developments in time — that all I was interested in was seeing that he received the recognition which was rightfully his! It made nor difference. The doors were closed.

It was now James Johnson's turn to promote commerce in Murray by advancing Nathan's legend, and he did the job admirably. Johnson had worked at WNBS in several capacities so he knew something about the radio business and its technology. He spent more than a year gathering Stubblefield memorabilia and convinced Vernon Stubblefield to give the Chamber all of Nathan's old photographs, documents, and personal papers. This bequest was fortunate, both for Johnson's project and for the conservation of the artifacts themselves. They had been stored in boxes, deteriorating in the damp basement at Vernon's drugstore. Johnson carefully cataloged, mounted, and preserved each item for posterity and for his own Stubblefield scheme.

The Kentucky Broadcasters Association invited Johnson to give a presentation on Nathan to its May 1961 convention at Louisville. He brought with him Nathan's 1908 US and Canadian patents for the wireless telephone, many photographs from 1902 and 1908, copies of newspaper accounts of the 1902 demonstrations, and Photostats of original documents. All of these exhibits were carefully and conveniently mounted in scrapbooks and available for all to peruse. His speech was a milestone in Stubblefield lore. Sentimental, energetic, and chock full of inaccuracies and glittering generalities, it was entertaining and effective. He claimed that by 1880, Nathan was already studying James Clerk Maxwell's theories of electromagnetism "and became thoroughly familiar with the Hertzian Wave [Hertz himself did not get this far until 1888.]. At the time Nathan B. Stubblefield was 20 years old and Marconi — a SIX YEAR OLD CHILD — was bouncing around his father's Italian Villa in rompers!"

James Johnson with Stubblefield's 1908 wireless telephone patent, part of the display he prepared for the Kentucky Broadcasters Association, 1961

The speech served Johnson's purpose perfectly. His efforts got statewide and national publicity for Murray, and the Kentucky Broadcasters Association presented him with a plaque recognizing Nathan as the true inventor of radio. Johnson described the experience:

> I have had surprising success in my publicity efforts since the story release on May 18, 1961. Although I had anticipated at least six more months of research, I could not pass up the opportunity to prove my claims to the Kentucky Broadcasters. Representatives of the companies involved, Marconi, National Gage [sic] and Wire, RCA and NBC were present, and although challenged, none offered any argument.

Subsequently, the Johnson speech became the official Stubblefield story, replacing the 1948 brochure, distributed by the Chamber to anyone who inquired about Nathan's exploits. The entire text is reprinted as AppendixE. But Johnson tempered his pronouncements somewhat and added to the confusion in an attempt to forge a distinction between Nathan, Marconi and other early radio pioneers. He wrote to Joseph Baudino at Westinghouse:

> In eighteen months of research, I have never found an instance where anyone of authority has challenged a single point relative to our claims. I do not have any desire to detract from the great work of Marconi, Poulsen, Tessenden [sic], DeForrest, or others who worked so hard to give the world Radio Broadcast as we know it today. I will insist that Nathan B. Stubblefield be given the credit he so richly deserves. Marconi's work was along one line, Stubblefield another. The results gave us two new forms of communication, not one. One was voice transmission, the other a coded electrical signal.

Johnson reiterated this obscure distinction to E.F.

Wright, who wrote asking for background on a magazine feature:

> I have not read the published story in *Electronics Illustrated*, even though I edited and approved the draft of the story several weeks prior to its publication date. I do not approve of the title "America's Own Marconi" for several reasons. We do not wish to detract from the wonderful work of Marconi in any manner, even though a great share of his credit should have gone to Clerk Maxwell, 1864, and Rudolf [sic] Hertz, 1880. Electrical impulse, or electromagnetic waves were well established by them, but were put to commercial use by Marconi. A second and more important reason, is the difference in their lines of work. Marconi used only electrical impulse and Stubblefield invented voice broadcasting.

A few years later, Johnson wrote to Emil Sveilis at United Press International. By this time, his references to differences between the two inventors had become vague.

> I can assure you that Nathan B. Stubblefield is truly the inventor of voice broadcast. In fact he holds the only patents, either American or Canadian, that perport [sic] to recognize the invention of radio broadcast. He was also issued patents for the electric storage battery, the basis of most of his discoveries. There is really no conflice [sic] between Stubblefield and Marconi, as their inventions were entirely different.

The next major Chamber effort on Nathan's behalf came in 1962, an attempt to convince Murray State College to name its science building for him. To support this scheme, the Chamber commissioned an operetta. With music by Paul Shahan and libretto by Lilian Lowry, *The Stubblefield Story* premiered in May 1962 at Murray State. In the opening scene, several teenage couples are at a barn dance late one evening in 1893. They see strange lights over at the Stubblefield place and start singing about Nathan.

> Willie:
> He sounds plum crazy, do you believe all that stuff about him talkin' through the air, without no wires anywhere. How can it be, sounds plum crazy! Wires in the air, well, I declare! Did you ever hear of such talk, such silly, silly talk about a darn fool dreamer such as he.
> Tom:
> Well, maybe ol' Nathan is a dreamer, but Sime told me that early one morn, when he was lookin' high and low for his cow, but couldn't find her, he heard a voice a-comin' down right out of the air, without no wires or strings nor no machines, or any of such things, but yet, it sounded just like ol' Nathan's voice, the voice that came out of the air and said! Sime, "is your cow out?" "can't you find your cow?" or anyhow, that's what Sime told me.
> Willie:
> Did he find the old cow? (Everyone laughs)
> Tom:
> Yep, but he didn't find no wires or string or any of those things.

Page from the original score, *The Stubblefield Story*, 1962

Pogue Library

NATHAN B. STUBBLEFIELD
INVENTOR OF RADIO
Born In Calloway County 1859
Died In Murray 1928

THE STUBBLEFIELD STORY

MUSIC-DRAMA IN FOUR SCENES

by

Composer ..... PAUL W. SHAHAN
Librettist .... LILLIAN LOWRY

May 23-24-25
Curtain Time 8:30 p.m.

Pogue Library

Program from *The Stubblefield Story*, 1962

The next scene takes place inside the Stubblefield house, also in 1893. Bernard is asking Nathan about his secret activities in the little black shack. Ada just wants Nathan to stop dreaming and bring in some firewood.

> Mrs. Stubblefield:
> Will you bring in the wood you promised or send Bernard. Let him do the chores you never do and quickly. It's almost suppertime.
> Nathan:
> Yes, yes, at once. Now listen to me boy. Would you like to see the work I do in that small shack? To work with me, keep my secrets?
> Bernard:
> Oh, yes, yes Father, tell me what to do.
> Nathan:
> First, you must understand. They laugh at me, they laugh at me and say that I'm a crazy fool; that all I do is scheme, and dream of unseen

things. But I must search for truth, alone. They do not know their world nor their tomorrows. Why even Shakespeare sensed in his own time the narrow limits of our little worlds when he said, "There are more things in heaven and earth, Horatio, than were dreamt of in your philosophy." Our neighbors are like that, our friends, most of them, good God, even my family, your mothers sees only foolishness in my search …
Bernard:
For what, Father, do you look for?
Nathan:
I look for the magic in the earth and the air around us, waiting to be known and used, waiting to be used and known, waiting to be known and used by any man patient enough …
Mrs. Stubblefield:
Nathan, I got the wood, I got it by myself. But if you want some potatoes for your supper, you better go fetch some.
Nathan:
Yes, yes, at once. I go at once. Did you know boy that in the earth and air there is power that we can use …
Mrs. Stubblefield:
Nathan, stop it. Stop it. Stop it. Stop it. Stop that silly talk and do something for a change. All this talk, talk, talk, is driving me insane. And, leave the boy alone. Don't make him a dreamer. Just talk, and tinker and jumble and clutter and brood your whole life through.

In the next scene, outside Nathan's home in 1898, his friend Duncan Holt comes by to see the latest inventions. After a ceremonial toast with elderberry wine (Nathan was never known to drink alcohol.), they duck into the shack for the experiment. There's a loud explosion, followed by flashing lights and smoke. Despondent, Nathan staggers out muttering: "Failure again. Defeat again. So little time. So little money to start over. So little time. So little heart."

The final scene takes place at the Court Square in Murray, New Years Day, 1902. The entire ensemble, plus barbers, a shoeshine boy, a bellhop and a chorus of 14 extras, is on hand to see Nathan's demonstration. They dance and sing their way to the train station to welcome the newspaper reporter from St. Louis (He actually didn't arrive until 10 days later.); then the fun begins. Rainey T. Wells is on hand as the Master of Ceremonies.

> Now we want the world to know that what we are about to witness is the result of the work of a Murray citizen, Nathan B. Stubblefield. (Cheers) It has been a source of great pleasure to me to help this great inventor obtain patents first, on a telephone in 1888. [At that time, Wells was 13 years old.] In 1898, after many disappointing experiments, he made and patented an electric storage battery. In 1892, friends, this astonishing inventor showed me in a private demonstration that he could make sound travel through the air without the aid of wires. (Murmurs) I know that is hard to believe, just as it is difficult to understand the dial telegraph system which he invented in 1890. Fact is, I don't understand them either. But we ordinary folks don't have to understand them to believe.

After the introduction, Nathan and Bernard man the telephone sets. They talk and whistle back and forth. The crowd is ecstatic. Everybody joins in the final chorus.

> He said he'd make a telephone that can work without no wires at all, that can carry sound way out in space, that can send a man's word, send a man's word many miles away. He proved that he could do it. He did just what he claimed. The world will know his story. The world will remember his name.

The curtain comes down at the moment of Nathan's greatest triumph — the day that, according to the Murray Chamber of Commerce, radio was born.

Murray State decided not to change the name of the science building, but this decision failed to deter the Chamber from its promotional campaign. On the whole, James Johnson's contribution to the Stubblefield legend was as important as L.J. Hortin's had been. But he was paid to promote commerce in Murray not recognition for Nathan. He did his job well. But the outside world was now paying attention to Stubblefield, using different data and analytical tools, and coming to contrary conclusions.

Under the Microscope

> There was the mysterious figure of Nathan B. Stubblefield, whose gravestone in Murray, Ky., names him the Father of Radio Broadcasting. He is said to have transmitted voice as early as 1892 and made public demonstrations in 1902, including one in Philadelphia and another near Washington. His U.S. Patent No. 887,357, obtained in 1908, was the subject of long litigation, which won him victories but no revenue. He died of starvation in 1928 in a shack in Kentucky.
>
> Erik Barnouw in *A Tower in Babel*, 1966

In December 1962, James Johnson received an enlightening letter from C.C. Powers of Owensboro, Kentucky. Powers mentioned a program, based on Johnson's speech, that he and two of Rainey Wells' grandsons had presented to a civic club there. Powers also commented on an article in a recent Texas Gas Company quarterly magazine *MCF* about Nathan. Stating that his "interest in this consists in seeing a fellow Kentuckian properly recognized," Powers then told Johnson of his recent communication with Dr. Charles Susskind, a professor at the College of Engineering of the University of California at Berkeley. Susskind, who was writing a history of electronics at the time, had just published an article "Popov and the Beginnings of Radiotelegraphy" that Powers read. He wrote:

> I was a bit disappointed that Stubblefield was not mentioned in the article. The article was so readable and interesting, as well as most serious and thorough, that I could not resist the temptation of bringing Stubblefield to the attention of Dr. Susskind. I am persuaded that he is in the unique position of being able, if so inclined, to sift the extant evidence and place Stubblefield in proper perspective. ... I took the liberty of sending him a complete copy of your KBA address.

Susskind replied and attached a draft of his manuscript "Observations of Electromagnetic Wave Radiation Before

Hertz." Johnson received copies of both items with Powers' letter. Susskind wrote:

> The question of whether he anticipated any of the better-known workers in the field would turn on whether he obtained transmission by induction (through air) and conduction (through earth or water), or by electromagnetic wave propagation. I am very skeptical of the statement made in the material you sent me that after the publication of the work of Maxwell and Hertz, Stubblefield "studied these theories religiously." Not many people were even aware of what Maxwell was doing, and most of those disagreed with his theory; the probability that a self-taught inventor could appreciate the work of the foremost mathematical physicist of the century is very slight. You may be interested in the enclosed manuscript, in which similar cases are discussed. You will note that even if Stubblefield did inadvertently produce an instrument that relied on electromagnetic wave propagation for transmission, he was not the first, since Dolbear demonstrated his device in 1882.

James Johnson filed the two letters and the manuscript away and never responded. Susskind's comments, however, were just the beginning of a body of research that attempted to place Nathan in proper perspective.

In the years after World War II, interest in the mass media evolved into a new academic discipline at colleges and universities across the US. Typically courses of study took one of two directions. The first was an extension of practical skills curricula, like the journalism program that L.J. Hortin established at Murray State University, to include radio and television broadcasting, advertising, and public relations. Many of these areas grew out of traditional subjects like English composition, public speaking, theatre, and business administration, and later became independent academic sequences. Within 30 years, mass communications were among the most popular undergraduate majors on many campuses.

The second thrust of mass media studies emerged primarily in graduate schools, with its roots in sociology, social psychology, cultural studies, and history. Soon this social science base expanded to include law, marketing, economics, aesthetics, critical theory, and a wide array of appropriate perspectives to study media content, business, audiences, and technology. By the late 1960s, mass media research produced a substantial body of social theory and hundreds of professors and graduate students willing to expand it with new research projects.

With all these academic types wielding microscopes and under the gun to complete dissertations or publications on which their careers depended, the new field quickly began to devour subject matter and research designs. To no surprise, Nathan Stubblefield's name began to pop up in general histories of broadcasting or in textbooks for introductory courses. Mostly, these authors treated Nathan anecdotally, as if they really didn't desire to explore his story but also didn't want other academics to think they were ignorant of it. Many of the writers, like Barnouw, contacted the Murray Chamber of Commerce so they relied heavily upon James Johnson's 1961 speech as a primary source. These accounts maintained Nathan's place within a chronology of technical developments. Paradoxically, however, by promulgating inaccuracies and coloring the story with adjectives like mysterious, peculiar, and eccentric, they also introduced Stubblefield folklore to a new and wider audience.

Perhaps the best example of a publication of this sort was an article that appeared in the October 10, 1970, issue of *TV Guide*. The author was Edward Lambert, a member of the faculty at the University of Missouri at Columbia. Titled "Let's hear it for Bernard Stubblefield! He was broadcasting's first entertainer.", the lively piece goes on to describe the performance on the Murray courthouse square, New Years Day, 1902.

> At a signal from his dad, Bernard Stubblefield took his place at the transmitter. First he began to talk.... Then Bernard began to whistle. Then he placed a harmonica to his lips and began to

> play. History's first radio program was following a variety format. Simultaneously everyone at the receivers heard him with remarkable distinctness. Sensationally, Bernard had proved the worth of his dad's invention.

Although he spun a good yarn, Lambert went into no technical detail. Except for the passage above, he very cautious with his terminology, referring to Nathan's experiments as "broadcasting," while subtly inserting the term "radio" to describe subsequent technology. For example, he writes:

> The crowd at Stubblefield's demonstration on New Years Day in 1902 had no way of realizing that they had been present at the dawn of broadcasting. Commercial radio's official debut was more than 18 years away, but the happenings that day offered a hint of things to come.

Readers probably ignored this fine distinction.

Elsewhere, and at the same time, the academic world was taking different approach to Nathan Stubblefield and coming to some contrary conclusions. These writers based their research on understanding the technologies that Nathan used within their historical context and on a fresh examination of primary sources, including the photographs and documents that L.J. Hortin, Vernon Stubblefield, and James Johnson had preserved.

The *Journal of Broadcasting*, a prestigious scholarly periodical, published two articles dealing with Stubblefield in 1970 and 1971. The first, "A Technological Survey of Broadcasting's 'Pre-History,' 1876-1920," was written by Elliot Sivowitch of the Division of Electricity and Nuclear Energy, the Museum of American History at the Smithsonian Institution. In a brief but factual chronology, Sivowitch described electricians, including Stubblefield, their experiments and inventions leading up to the development of radio broadcasting after World War I. As described in detail above, natural conduction, modulated light, static electricity, electromagnetic induction, and electromagnetic radiation were all used for wireless telecommunications in the late 19th century. In 1912, the US Navy began using the term *radio* to distinguish electromagnetic radiation wireless communications from all others. That basic definition and the term radio frequency (RF) to describe the waves involved are still in use. Nevertheless, in Great Britain and elsewhere the generic term *wireless* persisted, contributing to semantic confusion.

Sivowitch addressed that confusion. Of Nathan's 1908 patent for induction wireless, first tested publicly in 1892, he wrote:

> Now here is the critical point: most of the energy in the induction field is contained in the vicinity of the transmitting loop. The field, however, is varying at an audio-frequency, so far as this is concerned it obeys the same law as any varying field in space, regardless of frequency. Why isn't this radio? It turns out that we can determine from electromagnetic theory that there are three components of a varying electromagnetic field in space, one whose electric field varies inversely as the cube of the distance, $1/R^3$ (static field), one inversely as the square of the distance, $1/R^2$ (induction field), and one inversely as the distance, $1/R$ (radiation field). Some energy is *radiated* away from the antenna at any frequency, but at *low frequencies* (i.e., voice and music) most of the energy is confined to the vicinity of the wire. The induction field is the principal component of the Stubblefield system, and this has limited transmission range of something less than three miles.

Sivowitch also addressed the subject of natural conduction, the basic technology of Nathan's 1902 wireless telephone.

> ... [S]ystems involving conduction through the ground appeared over the next few decades and have been revived in modern times. These,

> however, were viable systems without any question, though only over limited distances. So far as our broadcasting story is concerned, however, ground conduction becomes intertwined with certain other related phenomena in the developing telephone technology. ... In 1877, a telephone "broadcast" was made from New York City to Saratoga Springs, New York. ... The musical programming was heard accidentally in both Providence and Boston due to electrical leakages between adjacent sets of wires on trunk lines north of New York City. Although conduction through the ground was the principal cause, *induction* through the air also was involved. Within both phenomena lay mechanisms for a new mode of communication. ... This line of development appealed to several late 19th century personalities. ... The crucial point to remember, however, is that the scientific base for induction-conduction communication was a natural outgrowth of conventional telegraph and telephone technology, and was not directly related to the Hertz-Marconi approach to wireless.

Two issues later, the *Journal of Broadcasting* published an article by Thomas Hoffer, an instructor and Ph.D. student in Communication Arts at the University of Wisconsin - Madison, titled "Nathan B. Stubblefield and His Wireless Telephone." At the outset, Hoffer stated:

> The purpose of this article is to document, from the fragmentary record remaining, what happened when Stubblefield experimented with batteries, coils and his strange "black box," and to provide an assessment of his work. The important question is whether his wireless telephone contained elements forming the first basis for wireless voice transmission, as it evolved into radio broadcasting; or, whether his was based on wireless "techniques" generally known by other experimenters of his time, and subsequently discarded in favor of other wireless theories.

Admitting that his evidence was "very sketchy," Hoffer summarized the events of Nathan's life and electrical experiments, based on a review of secondary sources and some documents in the archives at Murray State University. He also corresponded with Bernard Stubblefield, L.J. Hortin, and others. Although he failed to make distinctions between the two different wireless telephone designs and to grasp the details of the WCTA affair, the brief account was basically factual. Satisfied with the depth of his research, Hoffer concluded:

> The competence of persons testifying about Stubblefield's experiments cannot be challenged. But their competence about what was in Stubblefield's "black box" is certainly subject to question. Only Bernard, Stubblefield's son, had access to such information. Bernard Stubblefield has stated that his father's devices did not involve the generation of radio frequencies. ... Based on the available material, and the fact that wireless voice transmission evolved from the experiments of several persons widely separated by time and geography, it is clear that Nathan Stubblefield did not "invent radio broadcasting."

Two more extensive studies devoted to Stubblefield appeared in June, 1971. Both were doctoral dissertations written by residents of Murray. David Miller, at the University of Missouri - Columbia, titled his *The Role of the Independent Inventor in the Early Development of Electrical Technology*. Thomas Morgan, at Florida State University, called his *The Contribution of Nathan B. Stubblefield to the Invention of Wireless Voice Communication*. Taken together, these two manuscripts provided scholars with an organized and detailed database on Stubblefield. It was the first time that anyone had attempted that feat since L.J. Hortin and his journalism students in the year immediately following Nathan's demise.

The Morgan dissertation, the longer of the two, contains much information about Nathan that had never been published before. The author conducted a far ranging series of personal interviews with Nathan's surviving children, relatives, neighbors and acquaintances and collected a substantial quantity of oral history. He dug out the correspondence files for Stubblefield's patent applications and the history of the Wireless Telephone Company of America. He visited Bernard Stubblefield and got access to personal papers and other primary documents about Nathan. Morgan even went so far as to complete a rudimentary archeological search of Nathan's last home site, with minimal substantive results. He also acknowledged the attempts at recognition for Nathan's achievements and devoted a short chapter to this effort.

Pogue Library

Tom Morgan, second from right, returns Nathan's trunk and other artifacts donated by Bernard Stubbelfield, 1970 Murray State President Harry Sparks is second from left, and L.J. Hortin, far right.

Morgan's work, however is incomplete . Like Hoffer, he confused Nathan's 1892 induction wireless telephone, the basis for the 1908 patent, with the 1902 design that used natural conduction. Although Morgan touched on the subject of the Stubblefield's image since his death, he failed to realize its folkloric, even mythological dimensions and their impact on popular perceptions. Morgan provided limited historical context for the state of the art in wireless electrical experimentation and therefore reached the erroneous conclusion that: "The historical niche that belongs to [Stubblefield] is that of being the first man to successfully send and receive human voice without wires."

Nevertheless, Morgan evaluated Nathan's technical designs in minute detail by studying the photographs and patent illustrations. Then he compared these designs with contemporary radio technology. His conclusions regarding Stubblefield and radio were inescapable:

> Nathan B. Stubblefield is not the father of modern radio broadcasting. ... First, the waves that Stubblefield was generating are not today considered to be radio waves. ... Second, Stubblefield began with a system of induction and remained with it. ... Third, Stubblefield's system utilized an intermittent wave that was present only when one spoke into the microphone. Modern broadcasting is based on the process of modulating a signal upon a continuous high frequency carrier wave. ... Fourth, no one ever built on the Stubblefield system to elevate it in such a way as to make it the foundation of radio broadcasting.

By contrast, Miller used the Stubblefield story as a case study to explore the nature of inventors who worked independently of large companies. In doing so, he explored personality and lifestyle variables without the melodramatic narrative that categorized prior attempts to describe Stubblefield, the man. Miller's account of Nathan's early work with acoustic and electric wired telephones was a welcome and valuable addition. Because his was a study of invention, not just one inventor, he successfully placed Nathan within the mainstream of his contemporaries including Phelps, Edison, Marconi, Fessenden, and others. By implication, Miller made an important point often obscured by the subsequent promotional hype. Nathan was trying to invent a wireless telephone, not radio. In that effort, he succeeded, but only marginally.

Miller's dissertation, however, is too short and leaves us waiting for a substantive conclusion. Nonetheless, Miller restates in more sensitive, personal terms what others had determined:

> By the time Stubblefield's patent was granted in 1908, wireless telephony was already a practical reality. ... Due to the inefficiency of his wireless system in comparison to newer, more sophisticated high frequency "radiophones," the untutored Stubblefield was forced to witness as highly trained scientists and technicians brought to fruition many of his early predictions. ... Stubblefield's wireless telephone invention was only one of the many wireless inventions that were elbowed out of existence by later inventions and which enjoyed preeminence for only short periods of time. ... His two distinctly different wireless communication systems were only casually related to radiotelephony which was being developed at approximately the same time. ... [N]either of his systems was capable of producing the results that he envisioned.

Working independently and evaluating the same body of evidence, these four authors had come to similar conclusions. Nathan's place in history was approximately where George Clark at RCA, in his private memos during the 1930s said it would be -- an eccentric inventor who did the most he could with what he had. Now, those expert opinions were public and at odds with Nathan's popular image in his hometown.

Reevaluating the Legend

If the mounting evidence that Nathan's inventions did not lead to the technology we now call radio changed how the local community felt about their folk hero, it was not readily apparent. The City of Murray continued to issue wheel tax stickers imprinted with "Birthplace of Radio," and every car using the city streets carried one on the windshield. The Chamber of Commerce staged a revival of *The Stubblefield Story* in 1974 as an official Kentucky Bicentennial event. James Johnson kept on sending out copies of his speech, and L.J. Hortin, who returned to Murray State in 1967, persisted with a patient, low-key campaign for a grander memorial to Nathan.

By then, Murray State College had become a university. The student population grew at the fastest rate since the post-World War II years. The campus needed room to grow, so the administration began to buy up adjacent parcels of land, as they became available. Late in 1976, Murray State University made an offer on the property that had been Nathan's farm, located adjacent to the campus, and purchased the property the next year.

In 1980, L.J. floated a plan to construct a replica of Nathan's home on its original site and turn it into a museum. The Wrather Museum, less than a block away, would be responsible for management. Hortin estimated the cost at $100,000. Murray State Vice President Marshall Gordon agreed that the museum was a good idea but suggested that L.J. approach the Chamber of Commerce for the money. The plan ended there.

On the other hand, a non-profit company in Florida established the Nathan B. Stubblefield Foundation in 1979, but without Hortin's knowledge and not for the purpose that he intended. A wealthy philanthropist endowed a number of non-commercial community radio stations around the United States and set up private non-profit foundations to operate them. He named each foundation for a pioneer of wireless. The Nathan B. Stubblefield Foundation became the licensee of WMNF FM in Tampa, Florida.

Cast of *The Stubblefield Story*, 1974 revival performance

Pogue Library

Calloway County Bicentennial Committee
and Murray State University
Present

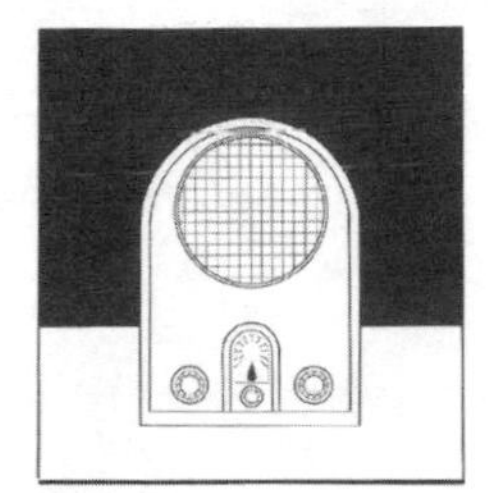

THE STUBBLEFIELD STORY

Libretto by Lillian Lowry
Music by Paul Shahan

Lovett Auditorium
Murray State University

Friday, April 26, 1974 8:00 PM
Saturday, April 27, 1974 8:00 PM

Pogue Library

Pogue Library

Program and publicity still from
The Stubblefield Story, 1974 revival

Embittered about the lack of public interest in the Stubblefield Memorial, Hortin nevertheless continued to write articles about Nathan for local newspapers in west Kentucky. Often, these were in the form of letters to the editor. They appeared regularly on the anniversaries of Nathan's birth, death, and major demonstrations and relied heavily on material from Hortin's 1930 feature in *Kentucky Progress* magazine. In addition, L.J. and James Johnson served as resources for many writers who produced features on Stubblefield.

Larry Kahaner, an associate editor of *73 Magazine* and an amateur radio enthusiast, published an article in the December, 1980 issue titled "Who Really Invented Radio? — the twisted tale of Nathan B. Stubblefield." This piece is different from others of the period in two respects. The author approached the subject with an open mind, and he bothered to come to Murray and sift through the evidence. Kahaner also had a thorough understanding of the technologies involved and contacts within the local amateur radio community. He posed the basic research questions:

> Who was that man with the bowler hat and handlebar moustache? And why, if he invented radio, has he been largely ignored outside of Murray? And why, if he had the willing financial backers for his invention, did he die a pauper, found locked inside his cabin outside of Murray where a pet cat seeking moisture had licked out his dried eyes? And why was it that the hundreds of articles written about Stubblefield, a Ph.D. thesis, and a play about his work failed to halt the controversy and contradictions surrounding this eccentric genius?

At the time the author did his research, the files at Murray State University's Pogue Library did not yet include the L.J. Hortin and Chamber of Commerce collections, so the author complained that most of the material was from magazine and newspaper articles and other second-hand sources. And he immediately encountered inaccuracies, confusion, and partiality.

> MSU's Stubblefield files contained materials (even from highly touted publications) that contradicted each other. I saw differences in simple items such as names, dates, spellings, and attribution. ... Each additional article I read only muddled the issue. ... In addition, it appeared that much of what has been written about Stubblefield was based on the research of two prominent Murray citizens who are less than unbiased about the role of the farmer/inventor in radio's early days.

Kahaner decided to base his article on what few primary sources were available, including Nathan's direct statements, reliable eyewitness accounts, the few photographs available to him, and the disclosure in the 1908 wireless telephone patent. Then he sought a range of opinions, including that of L.J. Hortin who he described as "distant and bitter about the whole affair." Hortin explained his frustration:

> I'm tired of people making fun of him [Stubblefield] and getting their information wrong. I've decided to put it all together and write a book. Pardon my vehemence, but I've been doing this for 50 years. I say he invented radio about 1890, but I don't think anyone really knows. ... Radio is a device that transmits and receives voice over considerable distance without connecting wires. Stubblefield invented, manufactured, and demonstrated such a device and did so before anyone else on the planet. That's my claim.

Kahaner also consulted James Johnson and got a copy of his 1961 speech. Then he talked with some local ham radio operators, including Riley Kaye, a retired radio engineer who worked for RCA and Western Electric, whose interpretation of the facts differed markedly from Hortin's:

> I think Stubblefield invented the induction tele-

> phone. He used loops above the ground. There appeared to be no carrier. He used audio frequencies, and that's where the challenge comes in. There is no proof that he used radiation. There's no proof that he used resonant circuits. That would be radio. Nobody can challenge that he didn't invent the wireless telephone and that he was the first to transmit voice without wires. He deserves a lot of credit and Murray can be proud of him.

Murray State professor Bill Call, another amateur radio buff, agreed but had an interesting perspective on the continuing controversy:

> It may have been magnetic induction, but you won't find that opinion around here much because it offends people. They want to believe he invented radio. On what I've seen, I don't believe he invented radio, but one thing almost everyone agrees on is that Stubblefield was a genius.

In the end, Kahaner returns to Nathan's 1908 patent and concludes that this device was not the forerunner of radio.

Most writers of this era, however, chose to disregard the convincing technical evidence about Nathan's inventions, preferring instead to extend the Stubblefield mythology. In truth, this option leads to more colorful prose. Brooke Kenvin, writing in *Television International Magazine* in 1987, described Nathan's big day on the Murray Court Square:

> It was a pleasant summer afternoon in Murray, Kentucky, in 1892. A crowd of prospective customers and investors had gathered around the courthouse lawn, unprepared for the historic event they were about to witness.
>
> In droves they came to see a fellow farmer (and telephone installer) named Nathan Stubblefield,

> who claimed he had a way to send music and human voice through the air *without* the use of wires. They watched in doubtful silence as the young man placed on the grass two two-foot square boxes about 200 feet apart and not connected in any way. Each box contained a telephone — and as Stubblefield and his son talked to each other from opposite ends of the lawn their voices could be clearly heard by the curious crowd. But after the miraculous feat was successfully performed — right there before their eyes and ears — the crowd responded with hoots and snickers, quite oblivious to the magnitude of the phenomenon they had been privy to — *the first public broadcast.* Frustrated and angry, Stubblefield cursed his own stupidity in conducting his sales presentation before such "dolts." But as he packed up his equipment and was leaving, he picked up five new investors — one of whom was Senator Conn Linn of Kentucky.

Although Kenvin writes with a distinctive flair, there is very little truth in the passage above. In 1892 as previously noted, Bernard Stubblefield was four, maybe five years old and hardly capable of assisting his father at this demonstration. Conn Linn was not practicing law in Murray, or anywhere else, because he too was a child. Linn and four others were investors in Nathan's failed scheme to market his 1908 patent. There is no record of this public event other than the Edward Freeman article of 1939, but it sounds very much like Nathan's experiments closer to 1900, perhaps even the famous one on January 1, 1902.

The fabrication continues. Somehow, Nathan got involved with Abraham "Honest Abe" White, the stock promoter who swindled Lee de Forest. White and Antonio Grazioni, according to Kenvin, raised "several million dollars by selling state territorial radio station rights and watered stock to confused investors throughout the wireless world." Neither man was ever mentioned in connection with the perpetrators of the Wireless Telephone Company of America fraud.

How did Nathan fail to get credit for inventing radio? Kenvin explained it this way:

> In 1905, by the stroke of a writer's pen, Marconi's wireless telegraph was dubbed "radio" — and radio was born. Unfortunately, in 1906 Stubblefield was forced by the U.S. Patent Office to categorize his invention as a wireless telephone broadcasting/receiving unit. Thus causing radio — as we know it today — to become lost in the name-game shuffle.

As mentioned earlier, the United Stated Navy defined *radio* in 1912 to differentiate wireless by electromagnetic waves from wireless by the natural conduction and induction methods that Nathan employed. Under this definition which persists today, Marconi was the first person to market *radio* communications systems. No one forced Nathan to call his invention a wireless telephone. That was his choice and his objective all along. Although he remarked on the 1902 wireless telephone's broadcasting capabilities, he did not use the term radio until 1922, and then only in response to prodding from his cousin Vernon.

Rather than stress the point about radio, however, Kenvin concludes the article by reverting to the confusing "name-game shuffle." In the process, she makes dubious assertions about Nathan's use of technology and places his work in a historical continuum based on those claims:

> Who won the broadcasting battle? There's no doubt that Nathan B. Stubblefield's wireless transmitter did. First with his own spark transmitter to broadcast music and the human voice for 200 feet, then 15 miles with the help of de Forest's early diode vacuum tube and later the capacitor .... At the turn of the century, the Kentucky farmer was simply overshadowed by Marconi — Italy's national hero — and his system of transmitting Morse code.
>
> If you ever go to Murray, Kentucky, visit radio station WNBS (the call letters spell out Nathan B.

> Stubblefield's initials) — they'll play your favorite song in remembrance of *their* national hero and his system of transmitting music and the human voice. Incidentally, the English still refer to the modern-day radio itself as *"the wireless …"*

Another approach to the Stubblefield saga has been to extend the legend and have some fun with it. A good example is the essay "The Real Father of Radio" by Lorenzo Milam, in his 1988 book *Sex and Broadcasting*. It begins:

> Somewhere in the shadows of the early history of radio looms the mysterious figure of Nathan B. Stubblefield. Nathan B. Stubblefield? Nora Blatch? Reginald Fessenden? Professor Amos Dolbear? Where do they get those names?
>
> He was born in, grew up in, lived in, and died in Murray, Kentucky. The citizens of that miniscule town were affectionate towards their mad radio genius, and erected a monument to Stubblefield in 1930. They called him the Father of Radio. Stubblefield was poor, and a mystic. He was a mendicant and a martyr to his invention … convinced that everyone wanted to steal it from him.

Milam ends the piece: "In 1928, Nathan B, Stubblefield of Murray, Kentucky died at seventy of starvation … and too many visions."

A remarkable article during this period was the entry on Nathan, written by Dr. Ray Mofield, in the *Kentucky Encyclopedia*. A professor at Murray State University, Mofield had an illustrious career in commercial broadcasting, became a William S. Paley Scholar at Columbia University, and was a much-loved educator. He created the broadcasting major at Murray State and put campus radio station WKMS on the air. Mofield's account is factual and concise. The surprise comes at the end:

> Dr. [Thomas] Morgan, cited above concluded Stubblefield did not invent radio. I conclude dif-

> ferently. Thomas Alva Edison, the man with more patents to his credit than any person living or dead in any nation, said that DC was the only practical ways to use electricity. He was wrong about the development of AC and Westinghouse was correct but this does not destroy the genius of Edison. Some other great scientists have reached the same conclusion. One of these was Dr. Eliot Sivowitch of the Smithsonian Institution in Washington.

At that point, Mofield quotes the passage from Sivowitch's *Journal of Broadcasting* article included in the previous chapter. But he stops with the question "Why isn't this radio?" Sivowitch goes on to explain that Nathan's invention did not qualify because it, like Dolbear's, did not transmit and receive radio waves. Morgan made a similar point, but Mofield chose to omit that assertion as well. Instead of building an argument on fact and logic, he chose selectively to ignore evidence to the contrary and extend theStubblefield legend.

On the eve of the 100th anniversary of Nathan's famous "Hello Rainey" transmission, the picture of this man and his accomplishments was still out of focus. As Larry Kahaner put it: "When they send you to unravel the twisted tale of Nathan B. Stubblefield, you're bound to run into trouble."

A Centennial Celebration

I first got to know Nathan Stubblefield early in 1990. Although I had heard a bit of the folklore, I didn't know that much about Nathan or the technology that he used. So I consulted Larry Albert, my friend and television engineer. He liked the idea enough to devote some spare time to the research. By then, most everybody who knew Nathan and had any substantive recollections was dead. L.J. Hortin, James Johnson, and David Miller were still around Murray, and it only took a phone call or two to scare up Tom Morgan in Florida. Most of the important archival documents, photographs, and memorabilia collected by Hortin, Johnson, and Vernon Stubblefield were stored at either the Pogue Library or the Wrather West Kentucky Museum on the Murray State campus. So we had plenty of stuff to wade through.

We proceeded with two objectives – to learn what Nathan did and to understand the state of the art of wireless electric communications in his era. With the Miller and Morgan dissertations and the work of other scholars as guides, we completed that phase within a few months. We decided that the best way to explain to a late 20th century audience what Nathan had accomplished would be to build authentic working replicas of his two wireless telephone systems. That way, people could see how they worked and decide for themselves what his place in history should rightfully be. Public demonstrations would also be a fun way to celebrate the centennial of the "Hello, Rainey" event.

We drafted a proposal and received initial funds from the College of Fine Arts and Communications at Murray State to buy the supplies and equipment that we needed. With the idea of creating a traveling exhibit, we began to look at various venues, like professional and academic conferences, that might be receptive to a demonstration.

By early 1991, we were well on our way. We had successfully tested models of both the induction and natural conduction systems, and found a source of replica telephone earpieces and mouthpieces. Because the period antiques were both fragile and expensive, we decided that the reproductions made more sense. Likewise, we opted to use 6-volt lantern batteries, rather than antique telephone batteries, as a power supply. There was no wooden box to conceal the circuitry. Unlike Nathan, we wanted people to see how these wireless telephones worked.

Then Keith Stubblefield showed up. A country music singer known professionally as Troy Cory, Keith was Oliver Jefferson Stubblefield's adopted son and hence Nathan's grandson. All his life, he had heard about how his grandfather invented radio but never got any credit for it. In 1972, his recording company, Cinema Prize Records of Universal City, California, started "an advertising and publicity campaign to build his name and to give credit where it's due to the inventor." (It's unclear whether "his" referred to Nathan or Troy Cory.) The blitz began with a press release entitled "The Greatest Broadcast Swindle, " compiled from sources including the books *Stranger Than Science* and the *World Almanac*, the US patent Office, and interviews with Troy Cory, who made the following rambling claim:

> "I think AMERICA," continues Cory, "has been cheated! The plundering anti-american [sic] position leading European controlled American newspapers, encyclopedia, and almanacs, had taken at the turn of the century by ignoring and omitting the Kentuckian's wireless feats from their publications and/or lists, not only undermined, but completely discridited [sic] and sold out one of America's Greatest Discoveries. WHY THE IDIOMS AND OMISSIONS? To get even? Ashamed of the yokel Inventor? To hide personal gains? National prestige? Or to Disguise a Stock Swindle That Took Place Between 1902 to 1912?"

Requests for information to Murray State University and to the Murray-Calloway County Chamber of Commerce brought Cory polite replies, copies of an L.J. Hortin article, and James Johnson's speech to the

Kentucky Broadcasters Association. Johnson later accused Cory of republishing the speech in his own name without attribution.

Nearly twenty years later, Troy Cory resurfaced in Murray with a centennial commemoration scheme of his own. Without the knowledge or permission of the University, Cory scheduled a press conference on March 9 at Murray State's Wrather West Kentucky Museum. Then invitations arrived with the following information:

> The morning Press Conference and Reception is to announce the preparation activities to celebrate, commemorate and honor the 100 [sic] anniversary of Broadcasting, It's [sic] Inventor, Nathan B. Stubblefield, and the birthplace of radio in 1892 - in Murray, Kentucky. Meet the staff members of Murray State University, the authorities on the history of radio and N.B. Stubblefield, Mr. and Mrs. Keith (Troy Cory) Stubblefield, members of the Stubblefield family, Bill Mack, the relief sculptor who will create the Stubblefield bust with his invention in relief, members of the staff and cast producing the N.B. Stubblefield television segment of Hawaiian Tropic's "Beauty From The Forbidden City," the Brooke Sisters, and the technical staff of MSU developing the replica of Stubblefield's original wireless telephone and his family home where it was invented.

In the days leading up to the press conference, Cory brought in a crew and shot some video in and around the campus. The high point of this production was the scene with scantily clad dancing girls on the tables in the normally calm reading room at the Pogue Library. Then Cory and his publicist, Chris Harris, met with MSU President Ronald Kurth, Development Director Chuck Ward, and others to consider his idea, a hodgepodge of many failed plans of the past. He wanted the University to reconstruct Nathan's home in its original location and turn it into a museum and also to name the new Industry and Technology Building for his grandfather. An alternative would be to turn the existing Wrather Museum into a Stubblefield commemorative site. His contribution to this plan was a strongbox of Stubblefield artifacts and memorabilia that he claimed was worth more than $100,000. In return, he wanted MSU to raise the rest of the money and commission him to write a multi-volume history of radio broadcasting, with a guaranteed fee of $100,000.

President Kurth was cool to the idea. He later told his staff that, after the press conference, Troy Cory was not to be allowed on the campus without his personal permission. With no announcement to make, Cory's press conference itself was an awkward, aimless affair. L.J. Hortin, one of the featured panelists, finally put on his hat and left.

Nonplussed, Cory returned home to California and arranged another press conference there. The *Los Angeles Times* reported the details April 12:

> Few schoolchildren would recognize the name Nathan Stubblefield, and, as a result, attorney Melvin Belli says he may sue book publishers on the behalf of misled youngsters. If you're asking who Nathan Stubblefield was, maybe you, too, can join in the plaintiffs' action.
>
> Stubblefield is the man who invented radio, according to his grandson, Troy Cory. At a news conference in Pasadena Thursday, Cory opened a recently recovered strongbox containing records which he said proved that Stubblefield, a Kentucky inventor, made his first broadcast in 1892. Thus, he would have been a few years before the breakthrough of Guglielmo Marconi, who many credit as radio's inventor (and the guy to blame for loud next-door neighbors).
>
> Cory, a Pasadena resident, was flanked by a Belli associate as well as publicist Chris Harris, who previously hosted the opening of Rudy Vallee's safe. (Historians are still mulling over the importance of those contents.) RCA and Westinghouse stole his kin's patents, Cory charged, though he admitted that it's too late to

> collect damages. But he warned book publishers that they better set the record straight.
>
> The grandson's real name, by the way, is Keith Stubblefield. He said he changed it to pursue a career as a singer. The name Stubblefield just doesn't seem to get much respect from anyone.

Cory then began a publicity tour that took him to the National Association of Broadcasters convention in Las Vegas in April, then to the Kentucky State Capitol, to the Kentucky Broadcasters Association, and to the Smithsonian Institution. The NAB refused to let him speak and made no offer to recognize Nathan. NAB general counsel Jeff Baumann said, "We do not see any relevance to the NAB convention, or to the association's purpose or name. It doesn't mean anything." Cory threatened to sue.

He returned to Murray in May and made a startling announcement. He was buying WNBS, the AM radio station whose call letters were his grandfather's initials, and its affiliated low power TV station. Many people in the broadcasting industry considered the purchase price of slightly more than $2 million to be excessive. For his part, Cory claimed that the broadcasting enterprise was only the first step in a grander plan for an entertainment complex. He planned to turn both stations into "superstations" beamed by satellite all around the world. Cory's initial broadcast would originate on a Stubblefield transmitter and be received in China, where he claimed to be a popular entertainer. Next would come the Stubblefield Museum of Radio Broadcasting, now relocated to the courthouse square adjacent to the radio and TV studios. The final phase included redeveloping the square as a tourist attraction complete with a high rise hotel and a moat encircling the courthouse.

With assistance from the owners of WNBS, Cory petitioned Kentucky Governor Wallace Wilkinson to proclaim 1992 as Nathan Stubblefield Year in the Commonwealth to celebrate the centennial of the first transmission. The governor acceded to the request and announced that Nathan was "the true inventor of radio" and Murray "the birth-place of radio."

The Kentucky Broadcasters Association, however, wouldn't go that far. At its May 23rd meeting in Pikeville, Kentucky, the KBA refused to reaffirm its 1961 statement. Instead it approved a resolution recognizing Nathan only for his "contribution to the early development of wireless transmission." Cory threatened to sue both the association and Francis Nash, who was writing a history of broadcasting in Kentucky. He also threatened legal action against Murray State University to enjoin the Stubblefield replica construction and confiscate "all Stubblefield artifacts, research and memorabilia - including a marker on campus indicating Stubblefield's homeplace." He wanted all the items shipped to his home in Pasadena where he would find a California university to complete the project and legitimate his claims.

His reception in Washington that July was not much better. By then, Cory was saying that a struggling radio museum in Dallas, Texas had agreed to name Nathan the true inventor of radio. He also hired Rick Wood, a Dallas engineer who specialized in radio frequency systems, to build working replicas of Nathan's wireless devices. Cory put on a demonstration on the Washington Mall for Elliot Sivowitch and others from the Electricity Section at the National Museum of American History. Sivowitch later said, diplomatically, "it didn't work very well." Robert Andrews reported the incident for the Associated Press:

> The grandson of Nathan B. Stubblefield, a Kentucky inventor and melon grower, confronted Smithsonian officials with the claim that Stubblefield invented the radio and Guglielmo Marconi wrongfully got the credit. The Smithsonian was unimpressed.
>
> After an hour of bickering in 90-degree heat Thursday before an audience of reporters and gawking tourists outside the National Museum of American History, Keith Stubblefield failed to win his grandfather's belated claim to fame. When

> all the shouting was over, Barney Finn, chief curator of the Smithsonian's museum's division of electricity and modern physics, was willing to concede only that "we have no problem with Stubblefield being an interesting and important historical figure" in the development of radio. ...
>
> Tempers grew short as Stubblefield and his publicist argued with Finn and Elliot N. Sivowitch, a museum specialist, about the significance of the Kentucky inventor's experiments. A shout of "liar" was heard at one point. Later, Finn turned to Stubblefield's publicist, who was carrying a tape recorder and said, "If we want to talk about history, that's one thing. If we're talking about a public relations campaign, that's another."
>
> "I have no problem with saying that Stubblefield transmitted with a wireless induction system," Sivowitch said, " but I wouldn't call it radio broadcasting."

By then, even the local media in Murray were fed up with Troy Cory's antics. In an August 9 editorial titled "Cory's campaign too hard to swallow," the *Murray Ledger & Times* said:

> Since beginning his crusade, Cory and Harris have antagonized and irritated Murray State University faculty and administration, Murray Mayor Bill Cherry, Dallas reporters, Washington DC reporters, Associated Press reporters, the Smithsonian Institution (especially their head of security) and everyone who has lived in Murray since 1928 and has "not done something to recognize him." You can now add *The Murray Ledger & Times* to the list.
>
> There is no mistaking that Stubblefield was indeed one of the first to transmit wireless sound and voice, and we are eager to see Stubblefield receive the recognition he deserves, whatever it should be. What is equally clear, however, is that

Cory and Harris have done more to take away from Stubblefield than they've given to him. They have been more interested in confronting than confiding.

In Murray, they called the residents "lazy." In Pikeville, Ky., they called people "liars." In Washington, DC, they called people "stupid." "We have to find an opponent; if we get beat, we look for another opponent," Cory once said about his campaign.

We have tried to ignore their method and concentrate on the message: Stubblefield's greatness. It has become harder and harder to do that. If they are ever to gain the trust of Murray and the historians who ultimately have to decide their debate — and it should be decided — they must cease attempts to cram their campaign down everyone's throats.

Forrest Pogue, left, and local radio personality Joe Pat James, right, listen to a replica of Stubblefield's induction wireless telephone, Murray, 1992

Meanwhile, Larry Albert and I proceeded with our historical replication. We finished the apparatus in February and tested it publicly in Murray first. One of the most gratifying moments at that event was seeing Forrest Pogue, one of L.J. Hortin's first journalism students, pick up the receiver, listen, and smile. That spring we took the Stubblefield Wireless Telephone demonstration to the American and Popular Culture Associations annual meeting in Louisville, then to the National Broadcasting Society convention in Arlington, VA. Elliot Sivowitch from the Smithsonian attended that one. He invited us back to the office to tour the storage area where we saw Alexander Graham Bell's original 1880 Photophone and what remains of Mahlon Loomis' 1866 wireless telegraph.

Stubblefield Centennial Exhibit, National Association of Broadcasters Radio Show, New Orleans, 1992

Our final trip was to the National Association of Broadcasters Radio Show that fall in New Orleans. The NAB gave us a booth on the exhibit floor, across from a display of the latest digital radio equipment. For three days, Larry and I demonstrated the 1892 Stubblefield Wireless Telephone and distributed literature to hundreds of radio broadcasters. After that, we returned to Murray and donated the equipment to the university. It has since become part of a permanent Nathan B. Stubblefield exhibit at the Wrather Jackson Purchase Museum.

Troy Cory was back in the news in October when the FCC approved the WNBS license transfer. He returned to Murray to enter the media business and compete with the *Murray Ledger & Times*.

Programming on the low-power TV channel began to suffer. Local news and sports, then syndicated and network programs soon disappeared from the schedule without notice. After a few weeks, the channel aired nothing but a simulcast of WNBS radio with a live camera aimed at the radio station control room. Cory first tried to force all employees to sign contracts as independent agents responsible for their own tax and social security payments. When that effort failed, he stopped paying them altogether. The employees were leaving in droves.

Cory returned to California in early December. Within a week, there was only one employee left to keep both stations on the air. On December 6, he called the Murray Police Department and asked for an officer to come to the studios. The policeman watched as the employee turned off both transmitters, switched off the lights, left, and locked the doors. WNBS, the radio station named for Cory's grandfather, that had identified Murray as the "Birthplace of Radio" every day since 1948, went off the air and stayed dark for a year and a half. The entire disaster took less than two months. It was the final event of Nathan's centennial celebration.

The next year, the city of Murray quietly redesigned the wheel tax sticker and removed the slogan "Birthplace of Radio."

Some Thoughts on Immortality

The Inventor and the Crank
By A Moneyless Man

Away back in the dark ages this story is laid
Just after we will suppose this world was made.
The inventor came in on a much needed mission
For the thing had been left in a messy condition.

This world had been left in the hands of an ox,
A man of no views, not enough orthodox.
Here inventor sailed in; he said it should run.
This Wizard, "The Crank," the son of a gun.

Who the ancient deluded, inventor did size,
Twixt devil and man but a poor compromise.
Inventor said little but thought more than twice,
Old Dennis could go, with his bag of advice.

Here genius was burning with fire of ambition,
Which Dennis mistook for the smoke of perdition.
Sad! Sad! was the life of the old time wizard.
His life must have been one continual blizzard.

But the time in its tide has spread this dominion,
Has burned old Dennis and his fossil opinion.
And the child of today these facts are now hearing
But for the "Crank" a darned small clearing.

Such men as Tom Edison, Franklin and his kind
Have portrayed for the world on the sands of time.

The wise in this age don't jump a conclusion,
Don't dub every thought save his own a delusion.
Don't think the inventor a freak or transgression.
He is the one who molds all worldly progression.

A "Crank" is a man with one idea sublimed,
To be pitied, not cussed, by all mankind.

Not a man of resources nor a man of refrain,
"Chant the winds when they blow o'er Gethsemane."

Nathan B. Stubblefield,
"The Man With The Irony Pen," 1886
Therefore I have contributed to the world's sum of knowledge just a little.

After installing the replica of Nathan's induction wireless telephone at the Wrather Museum, Larry Albert went back to his job as a TV engineer. I continued my research, picking up a few tidbits here and there to augment the archives at the Pogue Library and at the Calloway County Public Library. After L.J. Hortin died in 1994, he left all his private papers, including many priceless documents and piles of correspondence about Nathan, to Murray State University. I eagerly awaited the release of these files in 1998. Over the past few years, I've been fortunate to publish some magazine articles about Nathan in *Inc. Technology*, *Back Home in Kentucky*, *Paducah Life*, and elsewhere.

One of my favorite experiences began the day Tom Kimmel, one of my students, called. His older brother Brad, a successful video producer in Evansville, Indiana, had just finished a documentary about the Black Patch Tobacco Wars of the early 20th century for Kentucky Educational Television. KET liked the show and offered Brad a grant to produce another. He was thinking about Nathan as the subject.

Years ago, I was a TV producer and director. Nowadays, I teach production and writing. I was eager to retool my skills and to spread Nathan's story to a fresh audience. At first, I offered to help with research, but I ended up making arrangements for interviews with James Johnson, Forrest Pogue, David Miller, and others, scheduling location photography in Murray, and eventually writing the script. We even cast a couple of local actors as Nathan and Rainey Wells, found an apple orchard, and staged the "Hello, Rainey" bit. It was fun.

Crazy Nate and His Wireless Telephone: The Life and

Legend of Nathan B. Stubblefield aired for the first time on KET August 21, 1996. Brad was gracious enough to give me co-producer credit. Comments were generally favorable, and the show won a national award. But this program didn't tell the whole story due to time constraints. Our rough cut, which followed the script exactly, ran 44 minutes. KET had only allotted a half-hour. We probably had enough additional video to stretch it to a full hour, but most of it was talking heads — pretty dull stuff. So Brad had to cut about a third of the program. Most of what disappeared dealt with how Nathan became a folk hero after death. Correcting that omission was one objective for this project.

Keith "Troy Cory" Stubblefield continues his campaign to rewrite the history of radio and name his grandfather as its inventor. In 1993 he published the first of a planned multi-volume historical encyclopedia of the electronic media. Titled *The SMART-DAAF Boys: The Inventors of Radio and the Lifestyle of Nathan B. Stubblefield*, the book, according to the publisher:

> gives you a sense of the flow and flux of Murray, Kentucky, and soundness of its isolation, lassitude and entrapment. The story is written by members of the Stubblefield family, college professors, news correspondents, radio announcers and the secretaries for the attorney, it neither evades nor inflates. There's real shrewdness and compassion in the world's depiction of Stubblefield—but also a curious, lyrical sense of romance, intrigue and mystery. ... Troy Cory tells the Untold Story About Nathan B. Stubblefield, Radio/Television, the SMART-DAAF boys & the Evil " Bait-'n-Switch" Game Untold Story About Nathan B. Stubblefield, Radio/Television, the SMART-DAAF boys & the Evil " Bait-'n-Switch" Game. Clever Plastic Word - Produces Deception!

Cory's son, Scott B. Whitenack, Esq., (also known as Scott B. Stubblefield) has taken up the family cause as well, using Nathan's picture and heritage to promote his business, Radio Telephone Innovations. In an odd way, this marketing scheme is similar to one developed by Lucent Technologies, the AT&T spin-off company that owns the renowned Bell Laboratories. Lucent placed an illustration of Bell's Photophone on its stock certificate to stress the company's lengthy involvement in wireless and optical communications technologies. Whitenack's web page features the photograph of Nathan's Philadelphia demonstration with the large group of people. The caption makes the erroneous claim that both George Westinghouse and Nikola Tesla are in the picture, with the implication that they too stole Nathan's inventions.

Radio entrepreneur Erik Rhoads, publisher of *Radio Ink* magazine, mentions Stubblefield in his 1995 book *A Blast From the Past: A Pictorial History of Radio's First 75 years.* Working from a few newspaper clippings and a fertile imagination, Rhoads extended the legend a bit with the mistaken claim that " If you go to the town square of Murray, Kentucky, you'll find a statue of Stubblefield inscribed with the words "Murray, Kentucky — Birthplace of Radio." Despite a few other discrepancies, Rhoads briefly told the basic story but stated no opinion on Nathan's actual contribution to radio.

By contrast, writer Sharon Phillips Denslow made no pretense of trying to get the facts straight in her 1995 book *Radio Boy.* It's a "fictionalized account of inventor Nathan B. Stubblefield's childhood [that] invites young readers to investigate the history of science and technology, to envision what might not be impossible." *Radio Boy* is a typically benign children's story with beautiful illustrations that makes no claims about Nathan, other than that he was a bright boy. In an author's note at the end, however, Denslow writes, "his invention that transmitted voices through the air without wires was what we now call a radio."

One of Troy Cory's nemeses, Grayson, Kentucky radio station owner Francis Nash, wrote about Nathan in *Towers Over Kentucky: A History of Radio and Television in the Bluegrass State*, published in 1995. Nash, who relies heavily on the Miller and Morgan dissertations, takes issue with Cory's assertion about his grandfather:

> While attributing the invention of radio to a Kentuckian is more folklore than fact, Stubblefield certainly had visions about what such forms of communication might mean in the future. He is said to have proclaimed that one day his invention would be used to transmit news of every description. That prophetic message would prove more enduring than the limited technology of his invention.

Excerpts about Nathan from Nash's book also appeared in *Kentucky Living* and *Kentucky Explorer* magazines.

Another Kentucky writer who delved into the Stubblefield saga was Bill Cunningham, a Circuit Court Judge and historian of west Kentucky. He included a chapter on Nathan in the second edition of *Flames in the Wind: An Inspiring Collection of Stories About Courageous West Kentuckians* in 1997. In colorful, engaging language, he constructed an intriguing counterfactual argument:

> The life of Nathan Stubblefield punctuates a lingering sense of isolation and injustice which hovers just below the surface in west Kentucky. One cannot help but believe that if Stubblefield had been from Maryland, Brooklyn, or the outskirts of Boston, his fate would have been different. He would have been better protected by people with powerful connections, advertised and packaged by competent and honest hands. Instead, dazed and bewildered, he was exploited, fleeced, and then discarded back into the gray anonymity of remote and far flung west Kentucky.
>
> There is one irrefutable fact about Nathan Stubblefield. He had the courage to dream. Many people are fascinated by things which were. Nathan was fascinated by things which would be. He relentlessly pursued that dream, through poverty, heart wrenching loneliness, and

> public scorn and ridicule.

On the tough question of whether Nathan invented radio, Cunningham used his astute political judgment and demurred diplomatically. "Technically," he said, "it's a hard call."

Each month, the staff at the Pogue Library fields inquiries about Nathan, some of which result in publications. Leon Fletcher, a retired community college professor and amateur radio operator from California, wrote a piece for the *Monitoring Times*, October 1996, called "Okay, Class, The Inventor of Radio Was ..." Although he describes me incorrectly as "a staunch believer that Stubblefield was the first to invent radio," at least he spelled my name right. In the end, Fletcher fails to answer the question posed in the title.

A better piece of scholarship was the article in the May 1997 *Old Timer's Bulletin*, the official journal of the Antique Wireless Association. James Ryback, a professor of engineering at Mesa State College in Colorado, known for his work on Russian wireless inventor Aleksandr Popov, contributed a series of articles on "Forgotten Pioneers of Wireless," and the first was about Nathan. His conclusions resemble Tom Morgan's:

> Stubblefield's wireless telephones used audio frequency, not radio frequency, currents. He generated only audio frequency ground currents and induction fields. He likely did not propagate any "far- field" electromagnetic waves as did Marconi and others. The existence of these "far-field" waves is necessary if the term "radiotelephony" is to be applied appropriately. Stubblefield envisioned only short-range telephone communications, not long-distance wireless broadcasting. Nonetheless, Stubblefield deserves considerable credit for his pioneering work with wireless telephony.

In his amusing 1998 book *Raw Deal*, Ken Smith

devotes a chapter to Stubblefield. He depicts Nathan as a pitiable victim of circumstances:

> Nathan Stubblefield was an unschooled melon farmer from Murray, Kentucky, who invented the wireless telephone in 1902. Stubblefield has been mired in obscurity partly because of his humble origins, partly because his invention was flawed — but mostly because he was denied the opportunity to improve it after his Wall Street partners plundered his development company and left him bankrupt.
>
> Even with its imperfections, Stubblefield's telephone was a remarkable accomplishment. It could have ushered in the era of wireless communication in the 1920s, rather than the 1990s. Instead, Stubblefield — always suspicious of outsiders — was crushed by the betrayal of his business partners. He retreated into self-isolation, eventually starving to death in a pauper's shack. One of his final acts was to destroy every prototype of his invention.

Smith correctly categorizes the objective of Nathan's work as point-to-point communication rather than broadcasting. While it is true that Nathan left little or no equipment behind, it is more likely that he continually dismantled prototypes to recycle the parts. We have no way of knowing what his "final acts" were.

If you surf the World Wide Web with the keywords "Nathan B. Stubblefield" or "+Stubblefield +wireless," chances are that you will discover more than a thousand web pages that mention Nathan. Even after you cull out the duplicates, the false positives, and the ones that are only a name on a list, there will still be more than a hundred pages with some version of Nathan's tale. One of the more bizarre directions that Stubblefield lore took in the late 20th century stems not from interest in his wireless telephones but from a fascination with the Earth Cell Battery.

Apostles of metaphysics, pseudo-science, geomancy,

free energy theories, conspiracy theories, theosophy, and witchcraft now include Nathan among their pantheon. In his Web essay "Self-Preparation for the August [1999] Eclipse" in the David Icke E Magazine, Brian Desborough writes:

> During the latter half of the 19th century, several inventors (besides Tesla) discovered various methods for the harnessing of aetheric energy. While John Keeley was demonstrating his anti-gravity craft to the US Army and operating his implosion motor, Nathan Stubblefield was tapping into the planetary energy grid in order to provide electric lighting and heat for his home. The Illuminati maintains its dominance over humanity through its control of wealth, energy and most importantly, suppression of knowledge. If the discoveries of Keeley, Stubblefield, Tesla and other inventors had not been suppressed by the Illuminati, the world would have commercialized over-unity (free energy) devices and transmutation technology in the 19th century. The discoveries of these inventors necessitated that in order to maintain control, the Illuminati would have to withhold the true nature of aetheric energy from free-thinking members of the public.

In a footnote to the passage above, Desborough gives more details of what he believes Nathan did with his Earth Cell:

> Stubblefield obtained electricity with bifilar-wound coils consisting of adjacent windings of copper and iron wire. The coils were buried at the intersection of planetary aetheric energy grid points, preferably under the roots of oak trees. Sufficient energy was produced to power electric motors and arc lights.

Desborough's source is apparently *The VRIL Compendium*, an encyclopedia of obscure patents and experiments, published by Borderlands Science Research Foundation and compiled by Gerry Vassilatos. Nathan's work appears in *Volume VII, Dendritic Ground Systems* and *Volume VIII, VRIL and Ground Radio.* VRIL, a term coined by Sir Edward Bulwer Lytton in his 1871 novel *The Power of the Coming Race*, refers to natural electromagnetism, cosmic forces, or the "black radiant organismic energy of the earth" that may be controlled by a Philosopher's Stone or other magic devices. Alternately, New Age writers use VRIL to describe the language spoken on the lost continent of Atlantis.

Vassilatos has often written about Nathan. In his 1995 piece for the journal *Borderlands*, "Earth Energy and Vocal Radio: Nathan Stubblefield," he touches on the significance of natural power sources:

> Earth batteries are an unusual lost scientific effort having immense significance. Developed exclusively during Victorian times, the earth battery evidenced a unique earth phenomena by which it was possible to actually "draw out" electricity from the ground. The most notable earth battery patent is one which operated arc lamps by drawing "a constant electromotive force of commercial value" directly from the ground. In addition to this remarkable claim, a vocal radio system ... through the ground.

What follows is a unique picture of Nathan that reads more like a Tolkien fantasy than a history of technology:

> Who is Nathan Stubblefield, and why do most citizens in the state of Kentucky justifiably revere his name? A native of Murray, Kentucky, Nathan B. Stubblefield had a love for the lonely wooded areas on the outskirts of town. Certain spots in these woods were mysterious and possessed of a strange magic all their own. Few would seek the magic of these places, and learn its true and deep power. There the song is sweet and deep, and still.
>
> Vitalizing and sense-provocative, Stubblefield knew that specific locations could be unique and

> natural energy sources. Rock outcrops, evergreens, and flowing springs each registered as strong sensory attractants. Could it be that they were sensual attractants because they conducted and projected special energetic ground currents? Can it be that we are enthralled and drawn into certain spots because of their projective energy? Furthermore, what is the exact nature of this energy? Does it contain or exceed the qualities of electricity?

Vassilatos goes on to make unusual claims about Nathan's inventions and experiments. For example, he writes:

> In 1902, Stubblefield set up one of his sets in a "Mainstreet" upper office — in a hardware shop. From that point to his farm (some 6000 feet distant) he conducted continuous conversations with his son Bernard. Tapping with a pencil on his one-piece transceiver, Bernard was quickly heard in a loud, very clear voice. This transceiver was a carbon button placed in a tin snuff box. Speech and response were transacted through the self-same device, which acted as both microphone and loudspeaker. Cells were places downstairs from the office in the ground. They were never removed and never wore out, though operating twenty-four hours around the clock.

The passage above is a confusing mish-mash of anecdotes about at least three different inventions that occurred at different times. The bulk of the narrative describes Nathan's 1888 vibrating, acoustic telephone, which could operate around the clock because it needed no electricity. The rest is drawn from events 10 years later. Although the author requested and received some documents from the Stubblefield archives at Murray State University, he never did any research there. He lists no background sources, other than patents, and makes indefinite references to the origins of his information, then extrapolates wildly:

> I spoke with an academician who had the extreme privilege of speaking with Mr. Stubblefield's son Bernard Stubblefield. Bernard, by this time himself quite aged, told that his father's method in locating the "right spot" was deliberate. His father referred to the device [the Earth Cell] as indeed a receptive terminal and not a battery. Despite the insistence of the Patent Officers in calling the device a "battery," Stubblefield declared it to be an energy receiver ... a receptive cell for intercepting electrical ground waves. Its conductive ability somehow absorbs and directs enormous volumes of electrical energy. With this energy Nathan Stubblefield operated a score of arc lights at full brightness for twenty four hours a day.

A search of correspondence between Nathan, his lawyer, and the US Patent Office, stored at the National Archives, revealed that Nathan never referred to this device as an energy receiver. More often Vassilatos is completely vague in his identification of actual human beings who supplied evidence for his pronouncements:

> Witnesses convey that Mr. Stubblefield's batteries were usually buried at the roots of certain very old oak trees. From these sites it was possible for him to bring small arc lamps to their full candlepower. Tremendous amounts of energy are required for this expenditure of power. Not only was he remarkably able to draw such volumes of current from the ground reservoir for lamp lighting, but the power was available to him throughout the day.

In another essay, "An Introduction to the Mysteries of Ground Radio," available online at borderlands.com, Vassilatos makes similar allegations about Nathan's natural conduction wireless telephone, which the author states he invented at the age of 12:

> The very first vocal radio broadcast was

> engaged by Nathan B. Stubblefield (1872). Mr. Stubblefield employed special "earth cells" and long iron rods to transmit strong vocal signals "with great clarity." These signals traversed a mile or more of ground, a coordinated conduction wireless system providing wireless telephone service for a hardworking farm community. The Stubblefield Radio Method represents an essential technological mystery. His "earth cells" never wore out, never produced heat in their telephonic components, and provided "signal ready" power at any given instant of the day. Being neither activated or assisted by additional battery power, the system was fully operational around the clock.
>
> Later critics attempted the reduction of the Stubblefield Radio System to mere "subsoil conduction" mode of transmission, but remain completely unable to reproduce the performance to this day. Mr. Stubblefield repeatedly stated confidence that his Radio System was performing an act of modulation, not a transmission of signal power. The preexisting "electrical waves in the earth," he firmly stated, were the real carriers for his Wireless Telephone Exchange. The special "earth cells" were connective terminals, not power antennas; a means by which direct connection with the geomantic energy stratum was obtained.

As discussed in Chapters 4 and 5 above, Nathan's beliefs that electricity behaved like a fluid in the earth, air, and water around him and that he all he had to do was tap into this reservoir were mainstream physical science of his era. By the 1930s, however, most physicists rejected this view. As to the mysteries of the Stubblefield Radio System and its irreproducibility, I participated in building and demonstrating authentic replicas of it in 1992. Although Vassilatos occasionally stumbles upon a fact, for the most part his work is rife with wild speculation based on third-hand information, or even worse, tall tales.

Inventor? Crank? Wizard? How do you categorize Nathan B. Stubblefield? His life was undoubtedly "one continual blizzard," just as he foretold in his 1886 verses. But in the afterlife, Nathan's multifaceted image has sustained a rich folklore that resonates throughout the west Kentucky area where he lived. Some have proclaimed his genius in order to earn him his rightful place in the history of technology. Some have done so in the name of progress, and others for personal gain. Mostly though, those of us who get to know Nathan do so purely out of curiosity.

Whatever their authors' motivations, attention and inattention to detail, many accounts of Nathan's exploits exist. Over time, people tend to repeat the more colorful tales, true or not, and paint them with an expanded palette. To the casual reader, marginally inquisitive scholar, or World Wide Web surfer, all these stories have equal credence. And of course, when faced with the facts, most people choose to believe what they want to.

As much as I enjoy the image of Nathan the sorcerer searching the forest primeval for ancient gnarled enchanted oaks so that he might attach his magic earth cell to their cosmic taproots, I cannot allow this flight of fancy to interfere with my interpretation of the facts. Yet the New Age version, like the "Birthplace of Radio" campaign, the radio and stage shows, L.J. Hortin's lurid features, James Johnson's speech, and Troy Cory's rants that preceded it, probably contributes more to Stubblefield mythology than does his life story. Moreover, you cannot separate one from the other.

In a sense, the media created Nathan Stubblefield. It is unlikely that he could have known what he did about electricity, wired and wireless communications without reading the periodicals and journals available at the local newspaper office. Nathan constructed crude but effective advertising and publicity campaigns to generate attention from local media for his early inventions and telephone business. The Wireless Telephone Company of America was able to expand the exploitation. For about six months in 1902, Nathan got national attention and fleeting fame.

But then he lapsed from the public view until his obituary in 1928.

Next a group of young journalists and their mentor decided that Nathan's tale had news value and human interest. Their investigations and subsequent publications led to the first Stubblefield commemoration. Later, using newspapers, magazines, radio, and any other media platform available L.J. Hortin and Vernon Stubblefield spread the legend to wider audiences. Then the Murray Chamber of Commerce, the owners of radio station WNBS, and individuals like James Johnson, took up the Stubblefield cause as a ploy to focus attention on the town and promote industrial and commercial progress. Scholarly publications and general interest features followed, and soon a public, but bogus, debate over the invention of radio emerged. Nathan's grandson joined the fray, attempting to manipulate or control the media, rewrite the history of radio, and advance his own professional career. And so Nathan, the star of stage, screen, and the airwaves and the subject of books, articles, and press releases survived and remained an integral part of the popular culture of west Kentucky throughout the 20th century. Considering the obscurity of his late life and death, the creation of the Nathan B. Stubblefield legend is a remarkable media feat.

But relying solely on the media yields a superficial and often unreliable story. We get a list of accomplishments, details of a pathetic melodrama, and a wide range of opinions and interpretations. We know what Nathan did, but we don't know who he was. Even L.J. Hortin, David Miller, and Tom Morgan, the biographers who learned the most about him, failed to shed light on Nathan's personality and motivations. Constrained by a lack of evidence and by their own research objectives, they chose not to delve deeper than the basic facts.

L.J. Hortin started the ball rolling with his insightful journalism curriculum and his 1930 article "Murray, Kentucky, The Birthplace of Radio." While I disagree with his conclusion, I applaud Hortin's ingenuity. If only he had substituted the word *broadcasting* for *radio*, his argument would have been valid, and perhaps Nathan

Stubblefield and the community where he lived would enjoy more recognition. Likewise I am grateful to Vernon Stubblefield and James Johnson for their roles in preserving the scant evidence and primary sources about Nathan.

One exception to the factual narrative approach is the many attempts to explain or speculate on the last decade of Nathan's life. Morgan filled in some holes here, but most of this era is concealed by Nathan's hermit lifestyle and befuddled by the ignorance of eyewitnesses and by the perpetuation of local scuttlebutt. In casual conversation, itself an unusual event, Nathan remarks that he can light up a hillside. Soon, people are talking about him doing it, even though no one apparently saw the phenomena. Both L.J. Hortin and Tom Morgan spent years looking for eyewitnesses, and never found one. But by 1995, tiny lamps on the gossip tree grow into massive arc lights that required high voltages that Nathan must have drawn out of the nature with his magical earth cells. By tracing similar anecdotes through seventy years of rumor mills and sloppy journalism, it's easy to see how Stubblefield folklore flourished.

As tantalizing and mysterious as his late life was, the dearth of information about Nathan's formative years is a greater hindrance to our understanding of who he was, what he was trying to do, and why he lived as he did. He was intelligent, educated as well as most of his peers, and from a large but prominent family. Beyond that, we know little. What were his relationships with his mother, his father, his stepmother, and his siblings? What impact did the deaths of his parents have on him? Did he feel a competitive obligation to be successful like his father, grandfather, and brothers? What led to his dogged independence? So many questions with so few answers, and now no one is alive who can fill in the blanks.

We can only make minimal inferences. In their wills, both his father and older brother William went out of their way to keep anything of value out of Nathan's hands. His stepmother was a willing participant in the scheme to deed the farm to Nathan's children. They either thought he was profligate and a little crazy, or didn't like him very much.

We can't, however, determine how Nathan felt about them or their actions.

So we are left with a small body of documented facts. Nathan first chose farming, then invention as careers. His finest financial success came early, with the vibrating telephone business. And even though his best customers became his competitors and drove him out of business, Nathan was encouraged enough to try again with different objectives. He had four US patents, but failed to get one on his most important invention, the 1902 natural conduction wireless telephone. Neither that system nor his earlier induction wireless telephone, for which he received his last US patent in 1908, were the technological forerunners of radio. Moreover, he did not invent *the* wireless telephone, nor was he the *first* to use the technologies that he employed for that purpose.

Actually, Nathan was one of the *last* inventors to attempt wireless telephony by natural conduction or induction, his contemporaries having moved on to the more efficient electromagnetic waves. His inventions marked the end of an era rather than the beginning of one. So did his approach to the craft. By 1900, independent inventors were disappearing in favor of the corporate invention model created by Thomas Edison, George Westinghouse, and other industrialists. Invention, especially in the new fields of electricity and electronics, became too expensive and complex for a single person to afford or even understand. Part of Nathan's mystique, like Daniel Drawbaugh's before him, stems from the romantic, iconoclastic image proclaimed in the *St. Louis Post Dispatch* in 1902: "Kentucky Farmer Invents Wireless Telephone." In truth, he never had a chance to succeed once he left the confines of west Kentucky, where he knew more about wireless telephones than anyone, and moved into the larger technological environment where his credentials were marginal at best.

So the moment of his greatest notoriety, ironically, almost exactly coincided with his downfall. Nathan received wide acclaim for his demonstrations in Washington and Philadelphia in the spring of 1902. He signed what he believed was a legitimate contract with people he trusted to develop his invention commercially. What legal advice did he receive from Rainey Wells and others in this matter? As much public attention as Nathan got for the Murray, Washington, and Philadelphia displays, the one in New York turned out to be the most important. Following its failure in June 1902, his backers cast Nathan off in favor of an inventor who was more competent and cooperative. Calling his partners scoundrels, he returned to Murray to face friends who invested in his invention and ended up with worthless stock. Nathan never recovered financially from this fiasco and was unable to raise enough capital or get substantial publicity for his 1908 patent.

But Nathan Stubblefield was not a failure. He did exactly what he intended all along — invent a wireless telephone system. In fact, he created two of them. That in itself is remarkable considering that he worked alone, isolated from the intellectual ferment of an industrial laboratory, a research university, or a personal network of professional electricians. He also had the gift of perception so vital to process of invention, to see connections that the rest of us don't. He encountered the obstacle of a single channel environment in which all his wireless telephones shared one big party line. Nathan turned this defect into a potential asset by looking at wireless telephony from the consumer's point of view. He correctly foresaw, demonstrated, and publicly announced that broadcasting would be an important application for the future of such devices. Although he did not use the technology that would become radio, his wireless broadcasts were apparently the first, and his predictions about the mass media justifiably astute.

If anything in this volume has piqued your curiosity, most of the backgroundmaterial on Nathan and his legend are readily available at the Pogue Library at Murray State University. The Wrather West Kentucky Museumon campus has a smaller but impressive collection of Stubblefield artifacts. And Clayton Wells, a distant cousin of Rainey Wells, maintains a display of Stubblefield memorabilia at his Swamp Valley Museum in Denniston, Kentucky.

The folklore and the subculture of Stubblefield devotees are understandable. Nathan was a perfect subject. There was the mystery of his clandestine lifestyle and the experiments about which his own family knew little. Whenever he chose to demonstrate an invention, it was more like a magic show where the secrets remained hidden. Even today, most people do not understand the simple technologies that he employed. Although he favored privacy most of the time, in public Nathan was a tireless self-promoter, convinced of his own genius and the merit of his grand schemes. For a short time, he was both successful and famous, but he ultimately died as he had worked, alone. None of his inventions survived, but after death, his fame returned and endures. On the whole, Nathan did pretty well for a melon farmer from west Kentucky.

APPENDIX A

NATHAN STUBBLEFIELD'S US PATENTS

(No Model.)

N. B. STUBBLEFIELD.

LIGHTING DEVICE.

No. 329,864. Patented Nov. 3, 1885.

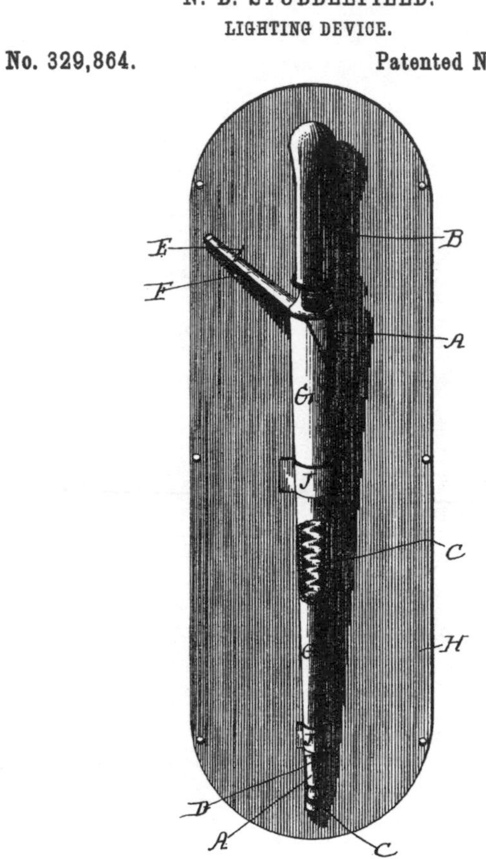

Witnesses:

W. F. Peterson
L C Linn

Inventor.

Nathan B. Stubblefield.

N. PETERS, Photo-Lithographer, Washington, D. C.

UNITED STATES PATENT OFFICE.

NATHAN B. STUBBLEFIELD, OF MURRAY, KENTUCKY.

LIGHTING DEVICE.

SPECIFICATION forming part of Letters Patent No. 329,864, dated November 3, 1885.

Application filed February 25, 1885. Serial No. 157,045. (No model.)

To all whom it may concern:

Be it known that I, NATHAN B. STUBBLEFIELD, a citizen of the United States, residing at Murray, in the county of Callaway and 5 State of Kentucky, have invented a new and useful Improvement in Coal-Oil-Lamp Lighters, of which the following is a description.

The object of this invention is to light lamps which have glass chimneys without the re- 10 moval of the latter.

To this end my invention consists in a tapering tube provided with a removable handle, a wick, a sheath fitting the tube, and a support for the sheath, constructed and com- 15 bined as hereinafter described and claimed, reference being had to the accompanying drawing, which is a front view, in perspective, of my invention, about three-quarters full size and partly broken away to expose the interior.

20 A represents a tin tube tapering from its handle B to its point, which is open. The handle is fitted to screw air-tight into the tube.

C represents a common cotton wick, which should be long enough to nearly fill the tube 25 when rammed or crowded in. This wick is to project from the point like a lamp-wick, and about a tea-spoonful of oil is to be placed in the tube with the wick. The air-tight fit of the handle prevents oil from escaping by 30 leakage.

D represents a groove through which the wick may be picked. This being dirty work for scissors, &c., I have provided a pick in the form of a spur, E, projecting permanently from 35 an arm or trough, F, which rises at one side of the tin sheath G.

The torch A B C is shown as inserted in the sheath, into which it fits closely, where it is kept when not in use, to prevent a disagreeable 40 smell and for sake of neatness. This sheath is removably secured to a back, H, which may be tacked to the wall at any suitable place. That it may be removed for the purpose of being cleaned, it is held in bands J, which are 45 permanently fixed to the back H.

When the wick is trimmed by the pick E, the charred ends, falling into the trough F, will slide down into the sheath, and there accumulate until it becomes necessary to clean the sheath. By removing the sheath from the 50 bands it may be inverted and jarred over the fire, when its contents will be disposed of.

The torch may be lighted at a fire or with a match, and, being long enough to reach down into the chimney of a lamp, it will light the 55 lamp without the necessity of removing the chimney. It may also be used for lighting gas and for other similar purposes.

What I claim as my invention, and desire to secure by Letters Patent, is— 60

1. The combination of the tapering tube A, the handle B, closely screwed therein, the back H, provided with bands J, and the sheath G, fitted to the said tapering tube and to the said bands, substantially as shown and described. 65

2. A sheath for lamp-lighters, having an open end and provided with a trough extended laterally from and opening into the sheath, and a spur or pick mounted on said trough, substantially as set forth. 70

3. A lighting device comprising a tube having an open end and a slot adjacent such end, and a sheath fitted to receive such tube, and provided with a lateral arm and a spur or pick mounted on said arm, substantially as set 75 forth.

4. The lighting device herein described, consisting of the hollow tube having an open end and a slot adjacent such end, a sheath fitted to receive said tube, the latter being re- 80 movable from the sheath, a trough extended from and opening into the sheath, and a spur or pick mounted on said trough, substantially as set forth.

NATHAN B. STUBBLEFIELD.

Attest:
S. L. HOLLAND,
L. C. LINN.

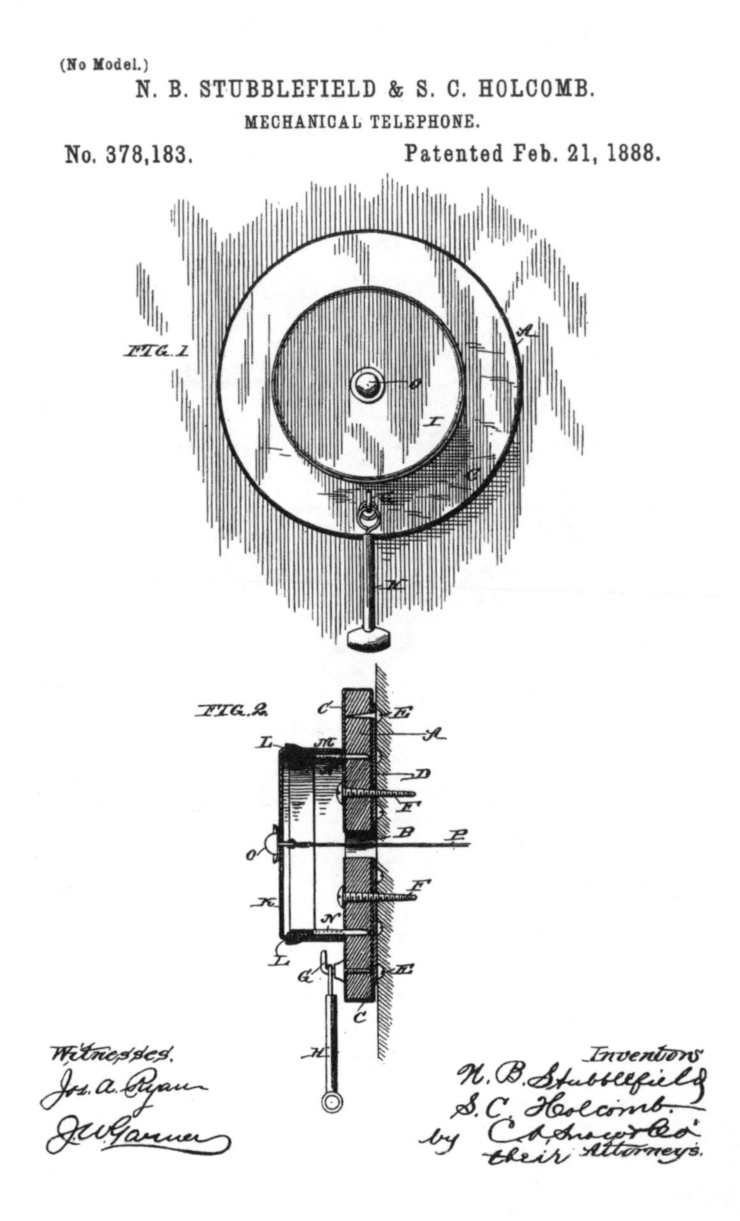

(No Model.)
N. B. STUBBLEFIELD & S. C. HOLCOMB.
MECHANICAL TELEPHONE.
No. 378,183.
Patented Feb. 21, 1888.
FIG. 1
FIG. 2.
Witnesses.
Inventors
N. B. Stubblefield
S. C. Holcomb.
by
their Attorneys.
N. PETERS. Photo-Lithographer, Washington, D. C.

UNITED STATES PATENT OFFICE.

NATHAN B. STUBBLEFIELD AND SAMUEL C. HOLCOMB, OF MURRAY, KENTUCKY.

MECHANICAL TELEPHONE.

SPECIFICATION forming part of Letters Patent No. 378,183, dated February 21, 1888.

Application filed February 19, 1887. Serial No. 228,231. (No model.)

To all whom it may concern:

Be it known that we, NATHAN B. STUBBLEFIELD and SAMUEL C. HOLCOMB, citizens of the United States, residing at Murray, in the county of Calloway and State of Kentucky, have invented a new and useful Improvement in Acoustic Telephones, of which the following is a specification.

Our invention relates to an improvement in acoustic telephones; and it consists in the peculiar construction and combination of devices, that will be more fully set forth hereinafter, and more particularly pointed out in the claims.

In the drawings, Figure 1 is a front elevation of an acoustic telephone embodying our improvements. Fig. 2 is a vertical central sectional view of the same.

A represents a base-board, which is made of wood, and is circular in shape and provided with a central opening, B. The face and edges of the board are covered with velvet cloth, as at C, the said cloth having its edges drawn over and upon the rear side of the base-board, and secured thereto by means of a circular sheet of tin, D, the latter having its edges secured to the base-board by means of headed nails E.

F represents a pair of screws which extend through the base-board, one above and the other below the central opening, B, the function of the said screws being to secure the telephone to a wall or other supporting object. From the face of the lower side of the base-board projects a hook, G, from which is suspended a mallet, H, that is used for calling the operator at the distant station.

I represents a circular drum which is made of sheet metal, preferably of tin. Over the outer side of the said drum is stretched a diaphragm, K, which is made of linen or other suitable cloth, and has its edge secured to the sides of the drum by means of cords L, which are wound and tied tightly around the drum, so as to surround the edge of the diaphragm, as shown at Fig. 2. The outer side of the drum is then covered with velvet cloth, as at M. From the inner edge of the drum project two or more securing-pins, N, which are adapted to enter corresponding openings made in the face of the base-board, and whereby the drum and the diaphragm are secured centrally to the base-board. The space between the base-board and the diaphragm and inclosed by the drum constitutes an air-chamber with which the opening B communicates.

O represents a button, the shank of which passes through an opening made in the center of the diaphragm, the button resting against the front side thereof. A wire, P, is attached to the shank of the said button and is drawn tightly, so as to tightly stretch the diaphragm, and extends through the opening B, the other end of the said wire being attached to a similar button in the diaphragm of a companion acoustic telephone located at the distant station.

In order to obtain the best results, we coat the diaphragm with copal varnish on both sides, which varnish, after it is dry, contributes materially to the resilience of the diaphragm and increases its sensibility, thereby enabling it to vibrate with maximum amplitude and mobility. The outer side of the diaphragm is then gilded or bronzed.

The simplicity of the telephone will commend itself to the public, while its advantages for talking over crooked or crossed lines will be seen at once.

Having thus described our invention, we claim—

In a mechanical telephone, the combination of the base-board A, the drum arranged on the base-board and having projecting pins N to enter the same, the cloth diaphragm stretched over the drum and secured thereto, the button on the outer side of the diaphragm and having the eye projecting through the center thereof, and the wire P, passed through an opening in the base-board and attached to the eye of the button, substantially as described.

In testimony that we claim the foregoing as our own we have hereto affixed our signatures in presence of two witnesses.

NATHAN B. STUBBLEFIELD.
SAMUEL C. HOLCOMB.

Witnesses:
OSCAR HOLT,
B. B. LINN.

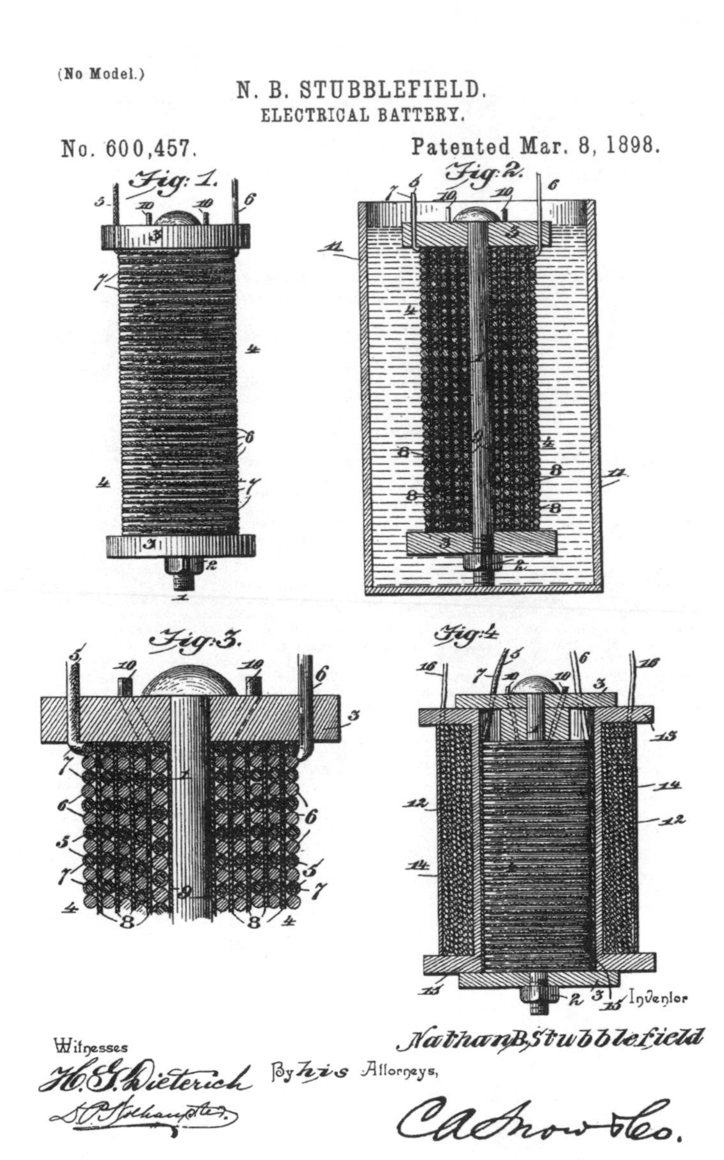
(No Model.)
N. B. STUBBLEFIELD.
ELECTRICAL BATTERY.
No. 600,457.
Patented Mar. 8, 1898.
Fig: 1.
Fig: 2.
Fig: 3.
Fig: 4.
Witnesses
H. G. Dieterich
By his Attorneys,
Inventor
Nathan B. Stubblefield
C. A. Snow & Co.
THE NORRIS PETERS CO., PHOTO-LITHO., WASHINGTON, D. C.

UNITED STATES PATENT OFFICE.

NATHAN B. STUBBLEFIELD, OF MURRAY, KENTUCKY, ASSIGNOR OF ONE-HALF TO WILLIAM G. LOVE, OF SAME PLACE.

ELECTRICAL BATTERY.

SPECIFICATION forming part of Letters Patent No. 600,457, dated March 8, 1898.

Application filed October 24, 1896. Serial No. 609,969. (No model.)

To all whom it may concern:

Be it known that I, NATHAN B. STUBBLEFIELD, a citizen of the United States, residing at Murray, in the county of Calloway and State of Kentucky, have invented a new and useful Electrical Battery, of which the following is a specification.

This invention relates to electrical batteries; and it has for its object to provide a novel and practical battery for generating electrical currents of sufficient force for practical use, and also providing means for generating not only a constant primary current, but also an induced momentary secondary current.

It is well known that if any voltaic couple be immersed in water or placed in moist earth the positive element of the couple will undergo a galvanic action of sufficient intensity to produce a current when the terminals of the couple are brought in contact, and this form of battery is commonly known as the "water" battery, usually employed for charging electrometers, but not capable of giving any considerable current owing to their great internal resistance. Now the principle involved in this class of batteries is utilized to some extent in carrying out the present invention, but I contemplate, in connection with water or moisture as the electrolyte, the use of a novel voltaic couple constructed in such a manner as to greatly multiply or increase the electrical output of ordinary voltaic cells, while at the same time producing in operation a magnetic field having a sufficiently strong inductive effect to induce a current in a solenoid or secondary coil.

To this end the invention contemplates a form of voltaic battery having magnetic induction properties of sufficient intensity, so as to be capable of utilization for practical purposes, and in the accomplishment of the results sought for the invention further provides a construction of battery capable of producing a current of practically constant electromotive force and being practically free of the rapid polarization common in all galvanic or voltaic batteries.

With these and many other objects in view the invention consists in the novel construction, combination, and arrangement of parts hereinafter more fully described, illustrated, and claimed.

In the drawings, Figure 1 is a side elevation of an electrical battery constructed in accordance with this invention. Fig. 2 is a central longitudinal sectional view of the battery, showing the same immersed in water as the electrolyte. Fig. 3 is an enlarged sectional view of a portion of the battery, showing more clearly the manner of winding the voltaic couple or, in other words, the wires comprising the couple. Fig. 4 is a vertical sectional view of the battery, shown modified for use with an induction-coil.

Referring to the accompanying drawings, the numeral 1 designates a soft-iron core-piece extending longitudinally of the entire battery and preferably in the form of a bolt having at one end a nut 2, which permits of the parts of the battery being readily assembled together and also quite as readily taken apart for the purposes of repair, as will be readily understood. The central longitudinally-arranged core-piece 1 of the battery has removably fitted on the opposite ends thereof the oppositely-located end heads 3, confining therebetween the magnetic coil-body 4 of the battery, said heads 3 being of wood or equivalent material. The coil-body 4 of the battery is compactly formed by closely-wound coils of a copper and iron wire 5 and 6, respectively, which wires form the electrodes of the voltaic couple, and while necessarily insulated from each other, so as to have no metallic contact, are preferably wound in the manner clearly illustrated in Fig. 3 of the drawings.

In the preferred winding of the wires 5 and 6 the copper wire 5 is incased in an insulating-covering 7, while the iron wire 6 is a bare or naked wire, so as to be more exposed to the action of the electrolyte and at the same time to intensify the magnetic field that is created and maintained within and around the coil-body 4, when the battery is in operation and producing an electrical current. While the iron wire 6 is preferably bare or naked for the reasons stated, this wire may also be insulated without destroying the operativeness of the battery, and in order to secure the best results the wires 5 and 6 are wound side by side in each coil or layer of the windings, as clearly shown in Fig. 3 of the drawings, so that in each coil or layer of the windings there will be alternate convolutions of the copper and iron wires forming 5 the voltaic couple, and it will of course be understood that there may be any number of separate coils or layers of the wires according to the required size and capacity of the battery. Each coil or layer of the windings 10 is separated from the adjacent coils or layers by an interposed layer of cloth or equivalent insulating material 8, and a similar layer of insulating material 9 also surrounds the longitudinal core-piece 1 to insulate from this core-15 piece the innermost coil or layer of the windings.

The terminals 10 of the copper and iron wires 5 and 6 are disconnected so as to preserve the character of the wires as the electrodes of the 20 voltaic couple; but the other or remaining terminals of the wires are brought in contact through the interposition of any electrical instrument or device with which they may be connected to cause the electric currents gen-25 erated in the coil-body 4 to flow through such instrument or device.

In the use of the battery constructed as described the same may be immersed in a cell or jar 11, containing water as the electrolyte; 30 but it is simply necessary to have the coil-body 4 moist to excite the necessary action for the production of a current in the couple, and it is also the contemplation of the invention to place the battery in moist earth, which 35 alone is sufficient to provide the necessary electrolytic influence for producing an electric current.

It has been found that by reason of winding the couple of copper and iron wires into 40 a coil-body the current traversing the windings of this body will produce a magnetic field within and around the body of sufficiently strong inductive effect for practical utilization by means of a solenoid or second-45 ary coil 12, as illustrated in Fig. 4 of the drawings.

The solenoid or secondary coil 12 is of an ordinary construction, comprising a wire closely wound into a coil of any desired size 50 on an ordinary spool 13 and incased within a protective covering 14 of mica, celluloid, or equivalent material. The spool 13 of the solenoid or secondary coil may be conveniently secured directly on the exterior of the coil-55 body 4 between the heads 3 with a suitable layer or wrapping of insulating material 15, interposed between the spool and the body 4, and the terminals 16 of the solenoid or secondary coil may be connected up with any 60 instrument usually operated by secondary currents—such, for instance, as a microphone-transmitter or telegraphic relay. The magnetic field produced by the current traversing the coil-body 4 induces a secondary 65 current in the solenoid or secondary coil 12, when the ordinary make and break of the primary current produced within the coil 4 is made between the terminals of said coil 4. It will therefore be seen that the construction of the battery illustrated in Fig. 4 is practi-70 cally a self-generating induction-coil, and it can be used for every purpose that a coil of this character is used, for as long as the coil-body 4 is wet or damp with moisture electric currents will be produced in the manner de-75 scribed. It will also be obvious that by reason of the magnetic inductive properties of the coil-body 4 the core-piece 1 will necessarily be magnetized while a current is going through the body 4, so that the battery 80 may be used as a self-generating electromagnet, if so desired, it being observed that to secure this result is simply required connecting the extended terminals of the wires 5 and 6 together after wetting or dampening the 85 coil-body.

Many other uses of the herein-described battery will suggest themselves to those skilled in the art, and I will have it understood that any changes in the form, proportion, and the 90 minor details of construction may be resorted to without departing from the principle or sacrificing any of the advantages of this invention.

Having thus described the invention, what 95 is claimed, and desired to be secured by Letters Patent, is—

1. A combined electrical battery and electromagnet, for use with water as an electrolyte, comprising a soft-iron core-piece, and a 100 voltaic couple of copper and iron wires insulated from each other and closely and compactly wound together in separate insulated layers to produce a solid coil-body surrounding the soft-iron core-piece, substantially as 105 set forth.

2. An electrical battery for use with water as an electrolyte comprising a voltaic couple of insulated copper wire and bare iron wire closely wound into a coil-body, substantially 110 as described.

3. An electrical battery for use with water as an electrolyte comprising a voltaic couple of insulated copper and bare iron wire wound side by side in separate insulated layers to 115 produce a coil-body, substantially as described.

4. An electrical battery, for use with water as an electrolyte, comprising a voltaic couple having its separate electrodes insulated from 120 each other and closely wound into a compact coil-body forming a self-generating primary coil when moistened and a solenoid or secondary coil fitted on the coil-body of the couple, substantially as set forth. 125

In testimony that I claim the foregoing as my own I have hereto affixed my signature in the presence of two witnesses.

NATHAN B. STUBBLEFIELD.

Witnesses:

JOHN H. SIGGERS,

W. B. HUDSON.

No. 887,357. PATENTED MAY 12, 1908.

N. B. STUBBLEFIELD.

WIRELESS TELEPHONE.

APPLICATION FILED APR. 5, 1907.

3 SHEETS—SHEET 3.

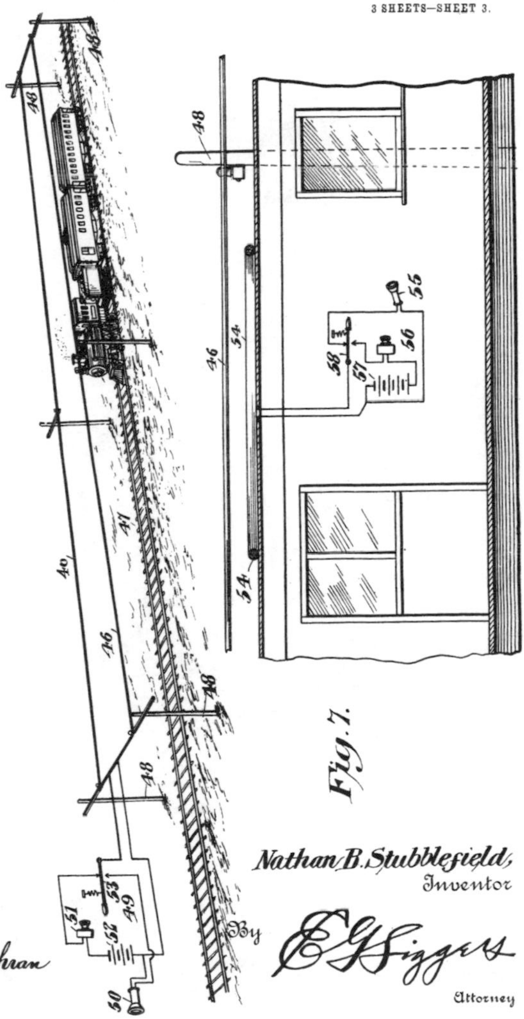

Fig. 6.

48 48 48 40 46 47 48 48 51 52 53 49 50

Fig. 7.

48 46 54 54 55 56 57 58

Witnesses

Jas. K. McCathran

B. G. Foster

Nathan B. Stubblefield, Inventor

By E. G. Siggers

Attorney

THE NORRIS PETERS CO., WASHINGTON, D. C.

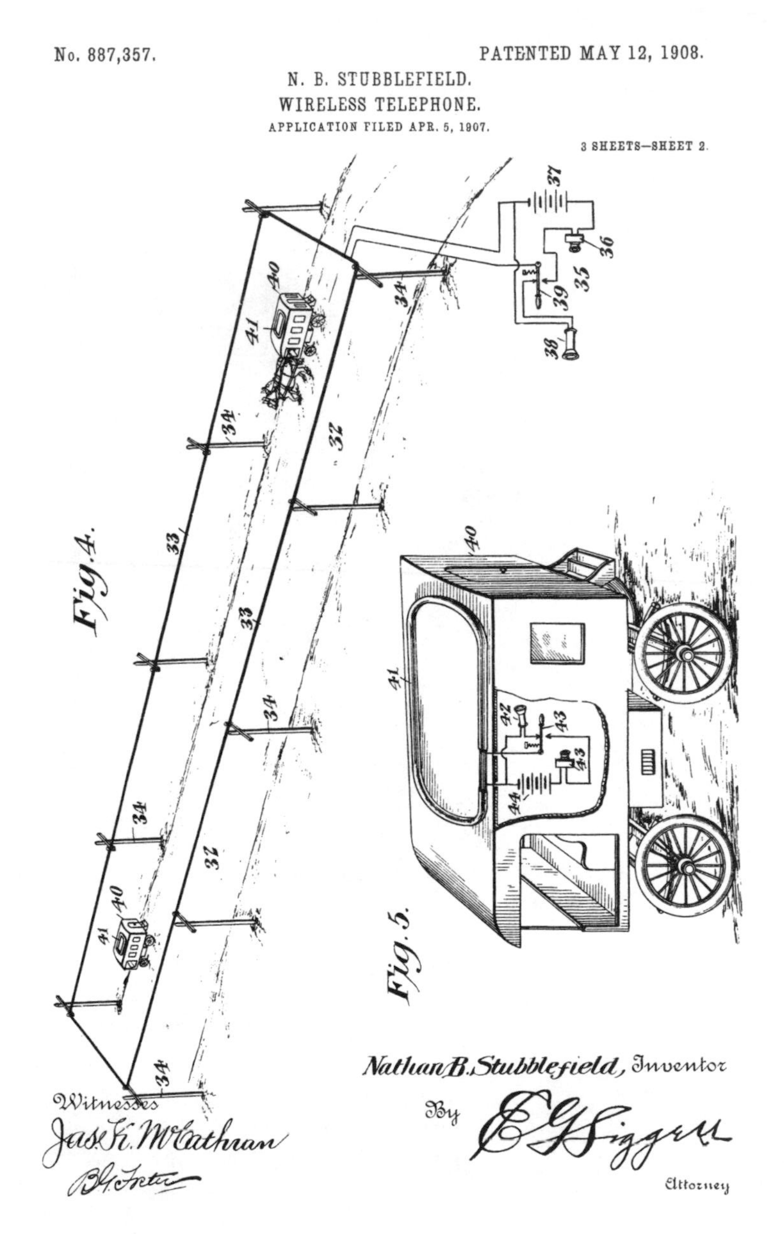
No. 887,357.
PATENTED MAY 12, 1908.
N. B. STUBBLEFIELD.
WIRELESS TELEPHONE.
APPLICATION FILED APR. 5, 1907.
3 SHEETS—SHEET 2.
Fig. 4.
Fig. 5.
Nathan B. Stubblefield, Inventor
By
Attorney
Witnesses

No. 887,357. PATENTED MAY 12, 1908.

N. B. STUBBLEFIELD.

WIRELESS TELEPHONE.

APPLICATION FILED APR. 5, 1907.

3 SHEETS—SHEET 1.

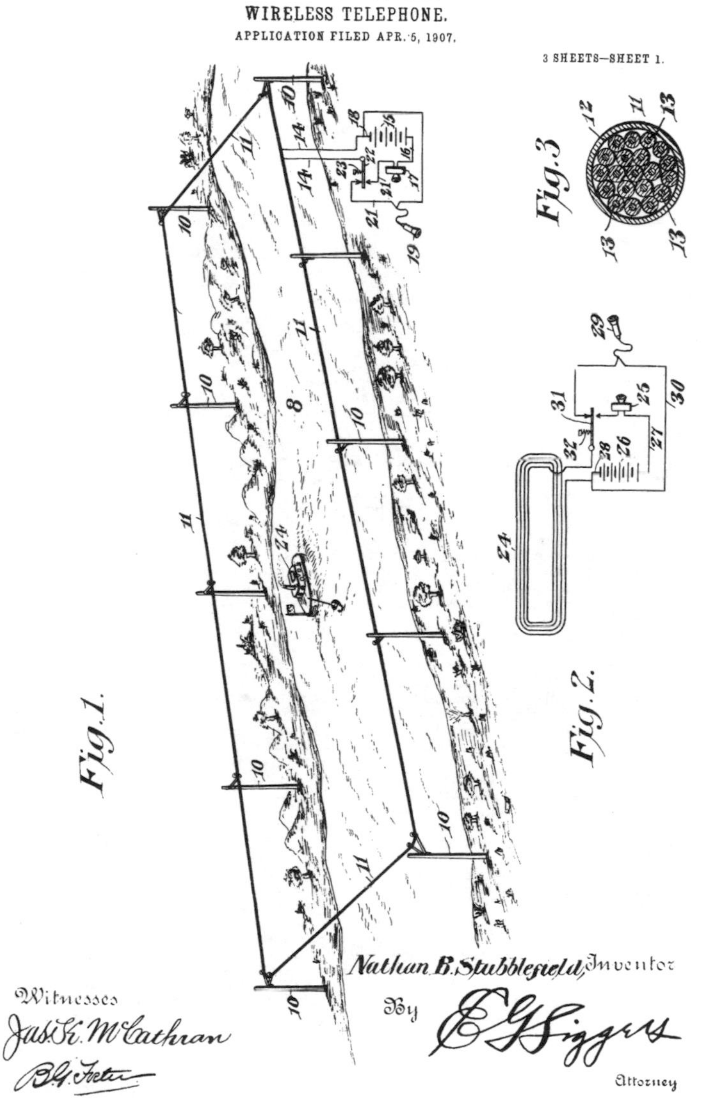

Witnesses
Jas. K. McCathran
B. G. Foster

UNITED STATES PATENT OFFICE.

NATHAN B. STUBBLEFIELD, OF MURRAY, KENTUCKY, ASSIGNOR OF TWELVE AND ONE-HALF ONE-HUNDREDTHS TO CONN LINN, FIVE ONE-HUNDREDTHS TO R. DOWNS, FIVE ONE-HUNDREDTHS TO B. F. SCHROADER, FIVE ONE-HUNDREDTHS TO GEORGE C. McLARIN, FIVE ONE-HUNDREDTHS TO JOHN P. McELRATH, TWO AND ONE-HALF ONE-HUNDREDTHS TO JEFF D. ROULETT, AND ONE-TWENTIETH TO SAMUEL E. BYNUM, ALL OF MURRAY, KENTUCKY.

WIRELESS TELEPHONE.

No. 887,357. Specification of Letters Patent. Patented May 12, 1908.

Application filed April 5, 1907. Serial No. 366,544.

To all whom it may concern:

Be it known that I, NATHAN B. STUBBLEFIELD, a citizen of the United States, residing at Murray, in the county of Calloway and 5 State of Kentucky, have invented a new and useful Wireless Telephone, of which the following is a specification.

The present invention relates to means for electrically transmitting signals from one 10 point to another without the use of connecting wires, and more particularly comprehending means for securing telephonic communication between moving vehicles and way stations.

15 The principal object of the invention is to provide simple and practical means of a novel nature whereby clear and audible communication can be established, said means being simple and of a character that will per- 20 mit certain of the station mechanisms to be small and compact.

In the accompanying drawings:—Figure 1 is a perspective view, showing means for establishing communication between a vessel 25 and a shore station. Fig. 2 is a diagrammatic view of the mechanism mounted on the boat. Fig. 3 is a cross sectional view on an enlarged scale of the shore coil. Fig. 4 is a perspective view of a road-way, showing a 30 system for establishing communication between road vehicles and a way-station, the latter being illustrated diagrammatically. Fig. 5 is a detail view of a vehicle equipped with one of the instruments, which is shown 35 diagrammatically. Fig. 6 is a perspective view showing the system applied to a railway for establishing communication between a moving train and a way-station. Fig. 7 is a sectional view through a car show- 40 ing in diagram the car mechanism illustrated in Fig. 6.

Similar reference numerals designate corresponding parts in all the figures of the drawings.

45 Referring to the embodiment illustrated in Figs. 1, 2 and 3, a water-way 8 is disclosed, upon which a vessel 9 operates. Surrounding the path of travel of the vessel, and preferably elevated on poles 10, is a coil 11 of 50 considerable magnitude. This coil, as shown in Fig. 3, consists of an outer casing 12, within which is placed a conducting wire comprising a plurality of convolutions 13, each of which is insulated from the other. The ter- 55 minals 14 of this coil extend to a suitable way-station, and at the station is located a powerful source of electrical energy 15, to which is connected by a suitable wire 16 an electrically operated transmitter 17. The 60 battery or other source of electricity has a connection 18 with one of the leads 14. A receiver 19 of the ordinary type has a connection with the same lead 14, to which the battery is connected, and both the receiver 65 and transmitter have connections 21 with the contacts of a switch 22. This switch has suitable means, as for instance, a spring 23, which normally maintains the receiver in circuit with the coil 11, as will be evident by 70 reference to Fig. 1, but if the switch is thrown to break the circuit, it will then cut in the source of electrical energy 15 and the transmitter 17.

75 An outfit similar to the above, is located on the vehicle or boat 9, but the coil 24 thereof, shown in Fig. 2, is much smaller. As further illustrated in said figure, the mechanism mounted on the boat, consists of a transmitter 25, and a battery or other source of 80 electrical energy 26 electrically connected, as shown at 27 and having a connection 28 with one of the leads of the coil. The receiver 29 also has a connection 30 with said lead. A switch 31 is connected to the other lead, and 85 is normally held in a position by a spring 32 to maintain a closed circuit through the receiver 29 and the coil, though it may be moved to cut out said receiver and close the circuit through the coil, the source of elec- 90 trical energy and the transmitter.

In this system, if it is desired to transmit from one station, as for instance, the shore-station, the switch 22 is moved downwardly to cut out the receiver and throw in the trans- 95 mitter and source of electrical energy, while the operator upon the boat or vehicle leaving the mechanism in the condition shown in Fig. 2, holds the receiver 29 to his ear. If therefore the operator at the shore-station 100 uses the transmitter in the ordinary manner, a varying current corresponding to that passing through the coil of great magnitude 11, will be induced in the coil 24, and the speech or other sounds will thus be transmitted to the operator on the boat. By reversing the arrangement, speech may be transmitted from the boat to the shore station.

The use of coils for both stations, each coil consisting of a plurality of convolutions has been found by experience to be of the utmost value, and furthermore experience has demonstrated that the employment of coils of different magnitudes is of great importance, for it has been found that while two small coils can be used to transmit but a short distance, if one large coil of the character set forth is employed, the other may be very small, and speech or sounds can be transmitted comparatively great distances from one to the other. These sounds are clearly audible.

The structure disclosed in Figs. 4 and 5 is of the same general character. A road-way 32 is disclosed surrounded by a coil 33 of great magnitude that is supported on suitable poles 34. The way-station 35 consists of a transmitter 36, a source of electrical energy 37 connected thereto, a receiver 38, and a switch 39, whereby the receiver or the transmitter and source of electrical energy can be thrown into circuit with the coil 33. The vehicles 40, which operate on the road-way, are provided with smaller coils 41 and instruments consisting of receivers 42, transmitters 43, sources of electrical energy 44 and switches 45 all arranged in the manner already described. In a system of this kind, it will be evident that the occupant of one vehicle may telephone to the home or way-station, and the message can be transmitted to another vehicle. Thus it will be evident that communication can be established between two moving vehicles or between a way-station and any vehicle desired which is within the range of the home- or way-station. The system is also capable of use in connection with railways, and in Figs. 6 and 7, such a system is disclosed in connection therewith. A comparatively great coil 46 is supported on opposite sides of the railway 47 by poles 48 and a station 49 has a receiver 50 and a transmitter 51, a source of electrical energy 52 and a switch 53, the last mentioned being employed for throwing either the receiver or the transmitter and source of electrical energy into closed circuit with the coil 46. One or more cars of a railway train is equipped with an outfit consisting of a coil 54, a receiver 55, a transmitter 56, a source of electrical energy 57, and a switch 58 for throwing either the receiver or the transmitter and source of electrical energy into circuit with the coil 54. It will be evident that the operation of these two last described systems are substantially the same as that first set forth, and no extended description thereof is believed to be necessary.

From the foregoing, it is thought that the construction, operation, and many advantages of the herein described invention will be apparent to those skilled in the art, without further description, and it will be understood that various changes in the size, shape, proportion, and minor details of construction, may be resorted to without departing from the spirit or sacrificing any of the advantages of the invention.

Having thus fully described my invention, what I claim as new, and desire to secure by Letters Patent, is:—

1. In a system of the character described, the combination with a vehicle, of a comparatively small coil of conducting material mounted thereon, electrical transmitting and receiving mechanism including a source of electrical energy connected to the small coil and carried by the vehicle, a stationary aerial coil of much greater magnitude than the small coil having its opposite stretches or sides extending along the opposite sides of the path of travel of the vehicle and elevated above the same and above the vehicle coil, and electrical transmitting and receiving mechanism connected to the greater coil and including a source of heavy electrical current.

2. In a system of the character described, the combination with a vehicle, of a coil of conducting material mounted thereon, electrical transmitting mechanism, a source of electrical energy connected thereto, receiving mechanism, means for connecting either the transmitting mechanism and source of electrical energy or the receiving mechanism to the coil, a stationary coil of greater magnitude surrounding the path of travel of the vehicle and comprising a plurality of convolutions of conducting material, the different convolutions being insulated one from the other, means for supporting the coil in an elevated position, electrical transmitting mechanism, a source of great electrical energy connected to said transmitting mechanism, electrical receiving mechanism, and means for electrically connecting either the transmitting mechanism and source of electrical energy or the receiving mechanism to the said coil of greater magnitude.

3. Means for communicating between a plurality of stations which consists of an aerial electrical coil of great magnitude, means for supporting the said coil, a station electrically connected to the great coil and comprising transmitting and receiving mechanism that includes a source of heavy electrical energy, and a plurality of other separate stations simultaneously in coacting relation with the aerial coil, each of said latter stations comprising a coil of conducting material spaced from but in coacting relation with said great coil and below the same, and transmitting and receiving mechanism connected to said other coil and including a source of electrical energy.

4. Means for communicating between a plurality of stations which consists of an aerial coil of conducting material of great magnitude, transmitting and receiving mech-
5 anism connected to said aerial coil and including a source of heavy electrical energy, a plurality of vehicles movable between the opposite sides or stretches of the great coil, coils carried by said vehicles and disposed
10 within the field of action of the aerial coil, and transmitting and receiving mechanism mounted on each vehicle and including a source of electrical energy.

In testimony, that I claim the foregoing as my own, I have hereto affixed my signa- 15
ture in the presence of two witnesses.

NATHAN B. STUBBLEFIELD.

Witnesses:
J. P. McElrath,
J. H. Coleman.

APPENDIX B

The following statement by Nathan Stubblefield probably originated as a press release. It first appeared publicly as part of the article on Stubblefield in the *St. Louis Post Dispatch* on January 12, 1902. Subsequently, it appeared in edited form in the prospectus for the Wireless Telephone Company of America in that year and was reprinted many times.

I have been working for this ten or twelve years. Long before I heard of Marconi's efforts, of the efforts of others, to solve the problem of the transmission of messages through space without wires, I began to think about it and work for it. This solution is not the result of an inspiration or the work of a minute. It is the climax of the labor of years, of days and nights of thought, of hundreds of hours of experimenting. Of course I worked along the line all others are working. The earth, the air, the water, all the universe, as we know it, is permeated with the remarkable fluid which we call electricity the most wonderful of God's gifts to the world, and capable of the most inestimable benefits when it is mastered by man. For years I have been trying to make the bare earth do the work of the wires. I know now that I have conquered it. The electrical fluid that permeates the earth carries the human voice, transmitted to it by my apparatus, with much more clarity and lucidity than it does over wires. I have solved the problem of telephoning without wires through the earth as Signor Marconi has of sending signals through space. But I can also telephone without wires through space as well as through the earth, because my medium is everywhere.

Beneath the surface of earth, as above it, there is electricity. No one knows how deep it extends or how high it goes. As one throws a pebble into a pond and agitates it into circles that grow and extend to every edge, the apparatus that I have invented agitates the electrical fluid in the earth. The voice projected into my transmitter agitates or vibrates the electricity in the earth, which extends beyond in every direction, and these vibrations produce the sounds in receivers tuned to convey them to the listening ear.

I can now telephone a mile or more without wires, and the expansion of my system is without limit. When the larger apparatus, on which I am working, is finished I will demonstrate that messages can be sent much further. The system can be developed until messages by voice can be sent and heard all over the country, to Europe, all over the world.

As to the practicability of my invention, all that I claim for it now that it is capable of sending simultaneous messages from a central distributing station over a very wide territory. For instance, any one having a receiving instrument, which would consist merely of a telephone receiver and a few feet of wire, and a signaling gong, could, upon being signaled by a transmitting station in Washington, or nearer, if advisable, be informed of weather news. Eventually, it will be used for the transmission of news of every description. I have as yet devised no method by whereby it can be used with privacy. Wherever there is a receiving station the signal and message may be heard simultaneously. Eventually I, or some one, will discover a method of tuning the transmitting and receiving instruments so that each will answer only to its mate.

I claim for my apparatus that it will work equally as well through air and water as it does through earth. That it will convey messages between land and sea, for instance, from houses to ships, from vessels in any part of the ocean to other vessels or to their owners on land if each carry my transmitters and receivers; it can be used by moving trains so that they may be spoken to between stations and thus prevent accidents. There is no conceivable position or station in which it may not be used. The all-enveloping electricity, the medium of carriage, insures that. The curvature of the earth means nothing to me. It will not deter messages sent by my apparatus.

I have shown you what my machine will do by grounding the wires. I will say it is not at all absolutely necessary to ground the wires. I can send messages with one wire in the ground, and the other in the air or with no wires at all.

The present method of grounding wires merely insures greater power in transmission. In fact my first and crude experiments were made without ground wires. I have sent messages by means of a cumbersome and incomplete machine through a brick wall and several other walls of lath and plaster without wires of any description.

Several years ago I invented an earth cell which derived enough electrical energy from the surrounding source to run a small motor continuously for two months and six days without being touched. There was enough energy in the motor to run a clock and other small pieces of machinery, or ring a large gong. This earth cell can be greatly magnified. Its discovery was the beginning of my experiments with wireless telephony. The earth cell merely buried in the ground and connected by the wires with the motor. The earth's electrical energy supplied the power. Its discovery was the beginning of my experiments with wireless telephony. The earth cell was merely buried in the ground and connected by wires with the motor. The earth's electrical currents supplied it with power.

The expense of my wireless telephone apparatus will not be great-nor greater than that used for ordinary telephoning, minus the present enormous cost of wiring.

APPENDIX C.

From *Scientific American*, May 24, 1902, page 363
THE LATEST ADVANCE IN WIRELESS TELEPHONY
By Waldon Fawcett

The latest and one of the most interesting systems of wireless communication with which experiments have recently been conducted is the invention of Nathan Stubblefield of Murray, KY, an electrical engineer who is the patentee of a number of devices both in this country and abroad. The Stubblefield system differs from that originated by Marconi in that utilization is made of the electrical currents of the earth instead of the etheral waves employed by the Italian inventor, and which by the way, it is now claimed, are less powerful and more susceptible to derangement by electrical disturbances than the currents found in the earth and water. In this new system, however, as in that formulated by Marconi, a series of vibrations is created, and what is known as the Hertizian electrical wave currents are used.

The key to the methods, which form the basis of all the systems of wireless telephony recently discovered - the fundamental principles of wireless telephony, as it were - was, discovered at Cambridge, Mass., in 1877 by Prof. Alexander Graham Bell, the inventor of the telephone system which bears his name. On the occasion mentioned Prof. Bell was experimenting to ascertain how slight a ground connection could be had with the telephone. Two pokers had been driven into the ground about fifty feet apart, and to these were attached two wires leading to an ordinary telephone receiver. Upon placing his ear to the receiver, Prof. Bell was surprised to hear quite distinctly the ticking of a clock, which after a time he was able to identify, by reason of certain peculiarities in the ticking, as that of the electrical timepiece at Cambridge University, the ground wire of which penetrated the earth at a point more than half a mile distant.

Some five years later Prof. Bell made rather extensive experiments along this same line of investigation at points on the Potomac River near Washington, but these tests were far from satisfactory. It was found on this occasion that musical sounds transmitted by the use of a "buzzer" could be heard distinctly four miles distant, but little success was attained in the matter of communicating the sound of the human voice. Meanwhile Sir William Preece, of England had undertaken experimental study of the subject of wireless telephony, and during an interval when cable communications between Isle of Wight and the mainland was suspended, succeeded in transmitting wireless messages to Queen Victoria at Osborne by means of the earth and water electrical currents.

Mr. Stubblefield's experiments with wireless telephony dated from his invention of an earth cell several years ago. This cell derived sufficient electrical energy from the ground in the vicinity of the spot where it was buried to run a small motor continuously for two months and six days without any attention whatever. Indeed, the electrical current was powerful enough to run a clock and several small pieces of machinery and to ring a loud gong. Mr. Stubblefield's first crude experiments looking to actual wireless transmission of the sound of the human voice were made without ground wires. Nevertheless, by means of a cumbersome and incomplete machine, without equipment of wires of any description, messages were transmitted through a brick wall and several walls of lath and plaster. As the development of the system progressed, the present method of grounding the wires was adopted, in order to insure greater power in transmission.

The apparatus which has been used in the most recent demonstration of the Stubblefield system, and which will be installed by the Gordon Telephone Company of Charleston, S.C. for the establishment of the telephonic communication between the city of Charleston and the sea islands lying of the coast of South Carolina, consists primarily of an ordinary receiver and transmitter and a pair of steel rods with bell-shaped attachments which are driven into the ground to a depth of several feet at any desired point, and which are connected by twenty or thirty feet of wire to the electrical apparatus proper. This latter consists of dry cells, a generator and an induction coil, and the apparatus used in most of the experiments thus far has been incased in a box twelve inches in length, eight inches wide and eighteen inches in height. This apparatus has demonstrated the capability of sending out a gong signal as well as transmitting voice messages, and this is, of course, of great importance in facilitating the opening of communication.

The most interesting tests of the Stubblefield system have been made on the Potomac River near Washington. During the land tests complete sentences, figures, and music were heard at a distance of several hundred hards, and conversation was a distinct as by the ordinary wire telephone. Persons, each carrying a receiver and transmitter with two steel rods, walking about at some distance from the stationary station were able to instantly open communication by thrusting the rods into the ground at any point. An even more remarkable test resulted in the maintenance of communication between a station on shore and a steamer anchored several hundred feet from the shore. Communication between the steamer and shore was opened by dropping the wires from the apparatus on board the vessel into the water at the stern of the boat. The sounds of a harmonica played on shore were distinctly heard in the three receivers attached to the apparatus on the steamer, and singing, the sound of the human voice counting numerals, and ordinary conversation were audible. In the first tests it was found that conversation were audible. In the first tests it was found that conversation was not always distinct, but this defect was remedied by the introduction of more powerful batteries. A very interesting feature brought out during the tests mentioned was found in the capability of this form of apparatus to send simultaneous messages from a central distributing station over a very wide territory.

Extensive experiments in wireless telephony have also been made by Prof. A. Fredrick Collins, an electrical engineer of Philadelphia, whose system differs only in minor details from that introduced by Mr. Stubblefield. In the Collins system, instead of utilizing steel rod, small zinc wire screens are buried in the earth, one at the sending and another at the receiving station. Single wire connects the screen with the transmitting and receiving apparatus mounted on a tripod immediately over the shallow hole in which the screen is stationed. With the Collins system communication has been maintained between various parts of a large modern office building, and messages have been transmitted without wires across the Delaware River at Philadelphia, a distance of over a mile.

APPENDIX D.

From *Kentucky Progress Magazine*, March 1930
Murray, Kentucky, Birthplace of Radio
By L.T. Horton (L.J. Hortin)

"Birthplace of Radio" is the honor that is being claimed by the citizens of Murray, Kentucky, who will dedicate a marker on the campus of Murray State Teachers College on March 28 in honor of Nathan Stubblefield, the first man to broadcast and receive the human voice without wires.

After proving to the world in 1902 that he could successfully transmit the voice by wireless, this Murray genius came home to die in a little hut near the city two years ago.

What price glory? Although he undoubtedly gave the world its greatest invention, the radio, he failed to get the honor due him. He wanted glory, for he wrote to his cousin Vernon Stubblefield: "You and I will yet add luster to the Stubblefield name. N.B.S."

Now the little tobacco and college town of Murray, Kentucky is trying to add that luster to his name by erecting in his memory a marker near the ruins of his old home. The tardy memorial will be dedicated March 28, 1930, exactly two years after his death.

"Hello, Rainey"-these were the first words conveyed by ether, the first radio message. To Dr. Rainey T. Wells, then a young attorney, Nathan Stubblefield in 1902 broadcast without the above message across a swampy Kentucky wood, now a beautiful campus. His only equipment consisted of a "crazy" box, some telephone equipment, two rods, and coils of wire. When news of the "crazy" Kentuckian's invention reached a Marconi-dazed world, "Eastern scientists and capitalists wearing diamonds as large as your thumb" flocked to the farm home of Nathan Stubblefield.

"I saw a written offer of $40,000.00 for a part interest in the invention," stated Dr. W.H. Mason, Murray surgeon, last week. Stubblfield refused half a million dollars because he thought his invention worth twice that sum.

On January 1, 1902, he demonstrated before a thousand people that the human voice could be broadcast and received without wires. A St. Louis Post-Dispatch reporter, in a full-page article of January 12, 1902, says:

"However undeveloped his system may be, Nathan B. Stubblefield, the farmer inventor of Kentucky, has assuredly discovered the principle of telephoning without wires. Today he gave the Sunday Post-Dispatch a practical demonstration of his ability to do this and discussed his discovery as frankly as his own interest and self protection would permit."

A soda keg first housed the invention that was to give him glory. Only Bernard, his 14 year-old son, was intrusted with the secret of the mysterious keg. With a shotgun he repulsed over-inquisitive visitors.

Invited by the leading scientists of the day, he went to Washington, D.C. On March 20, 1902, he demonstrated that his contrivance was practical. From the steam launch, "Bartholdi" on the Potomac River, he broadcast to scientists on the riverbank. In a framed photograph, made on this occasion, he wrote, "First Marine Wireless Telephone demonstrated in the world by Nathan Stubblefield. Four hour test cost twenty-five dollars."

On Decoration Day, 1902, Mr. Stubblefield proved to a critical audience of inventors, statesmen, business men, and newspaper reporters that his voice could be heard by wireless a mile distant from the transmitter. The Belmont Mansion and Fairmount Park at Philadelphia were the experiment grounds.

Fame was within his grasp. He obtained patents for his invention in England, United States, and Dominion of Canada. The number of his United States Patent is 887,357, dated May 12, 1908. The patent from the dominion of Canada is numbered and dated 114,737, October 20, 1908.

Radios were installed in automobiles for the first time not more than two years ago. But this unfortunate genius clearly anticipated such a modern luxury as early as 1908. In the original Canadian patent is a drawing made by Stubblefield of a "horseless carriage" with a broadcasting set, which he later called "radio." The same idea was to be used in trains and steam ships, the patent declares.

Marconi's name was linked with that of Stubblefield by Trumbull White in 1902 in a copyrighted book "The World's Progress." The article, headed "Telephoning without Wires," on page 297 states:

"Of very recent success are the experiments of Marconi with wireless telegraphy, an astounding and important advance over the ordinary system of telegraphy through wires. Now comes the announcement that an American inventor, unheralded and modest, has carried out successful experiments of telephoning and is able to transmit speech for great distances without wires."

To the St. Louis reporter Stubblefield related his knowledge of his contemporary, Marconi:

"I have been working for this, ten or twelve years, long before I heard of Marconi's efforts, or the efforts of others, to solve the problem of transmission of messages through space without wires. I have solved the problem of telephoning without wires through the earth, as Signor Marconi has sending signals through space. But I can also telephone without wires through space as well as through the earth, because my medium is everywhere.

As early as 1898 the Electrical World recognized him as an inventor of a "primary battery," which he later declared was "the bed rock of all my scientific research in radio today." He adds, "Dr. Pupin's coil, used in Trans-Atlantic cable telephone, came in part from this."

"He discovered a great law, built apparatus that would put the law into practical operation, made the apparatus work in practical demonstration of the law and principle-and in 1902 he forecast radio and all its branches, including broadcasting," wrote George K. Sargent, vice-president of Mutual Life in New York in an article published November 25, 1929.

Stubblefield predicted that he "or somebody else would discover a method of tuning the transmitting and receiving instruments so that each would answer only to its mate."

In a recently syndicated article, December 16, 1929, the NEA declares:

"The first two-way telephone conversations between a short point and a ship at sea were held the other day when persons in eastern cities talked to officers and passengers aboard the steamship Leviathan 200 miles out of New York. The historic event is pictured as H.A. Lafount, left, of the Federal Radio Commission, D.W. Chapman, president of the United States Lines, and William S. Gifford, right, president of the American Telephone and Telegraph Company, talked with Commodore H.A. Cunningham, master of the Leviathan, through the ether. Below is the short wave length telephone equipment in the deckhouse abroad the Leviathan. Here incoming and outgoing calls are handled at a charge of $7 a minute up to a distance of 1400 miles."

Although the obscure Kentucky farmer did not broadcast and receive telephone conversations for as far as 200 miles, he actually did give the first marine demonstration of radio-telephony 27 years before the "historic event" of the Leviathan.

At the peak of his life, he formulated plans for an industrial college-with himself as president. "Telephondelgreen, The Home of the Nathan Stubblefield Industrial School and Experiments in the Wireless Telephony" was to be the name of the institution. By some ironic twist of fate, his school is in ruins. On the same grounds is a million-dollar Kentucky college, established by Dr. R.T. Wells, the young attorney who first heard the inventor's wireless message.

Two years ago, the lifeless body of Nathan Stubblefield was found in a two-room shanty, crudely constructed by his own hands. His body, found 48 hours after death, was partly eaten by a hungry cat or by rats. Weakened by disease and partial starvation, the radio inventor had fallen and died-alone, except for a cat and a cow.

A hermit, partly insane, the wireless inventor had renounced his wife, children, and his relatives. Because the world had withheld the glory due him, he had forsaken his friends. Even the radio, instrument of his research, failed to mention his demise.

"When I want help, I'll call for it," he had curtly told his brother who had come to his assistance. His own daughter, wealthy, was driven back when she came from Mississippi to visit him.

The hut in which he lived is six miles north of Murray. Lined with cornstalks to keep out the cold, damp winds, the house was constructed at a cost of thirty-five cents, the people of that Almo neighborhood say. Coils of wire, pieces of nickel and steel, dust, and molded magazines on electricity are all that remain.

Why did this inventor fail to get the glory for which he paid all? With fame in his grasp, why did he come home from Philadelphia? Did someone steal his invention? Was he actually insane? Did family troubles cause his tragic failure? What was in the box, the secret of which he guarded so jealously?

From his patent it is apparent that he discovered the fundamental principles of broadcasting and receiving that have made the modern radio possible. Although he revealed to no one, except perhaps to his son, the contents of the box, the accompanying diagrams and his patent disclose the basis of his invention.

Supplied with sufficient electrical power from his "ground cell," Stubblefield discovered that electrical waves could be sent through the ether by using coils of wire.

"This coil," his patent declares, "consists of an outer casing, within which is placed a conducting wire comprising a plurality of convolutions, each of which is insulated from the other. The terminals extend to a station at which is located a powerful source of electrical energy, to which is connected by a suitable wire an electrically operated transmitter." This in brief constitutes his broadcasting station.

To a similar arrangement of coils, "on a smaller scale," the inventor attached his receiving apparatus, likewise connected with a "source of electrical energy." The remarkable part about his invention was that by reversing a switch, he could change his broadcasting station into receiving apparatus.

His early antennae and aerial consisted of wires which "terminated in steel rods, each of which was tapped with a hollow nickel-plated ball of iron, below which was an inverted metal cup." The wire entered the ball at the top and was attached to the rod. The rod, according to the St. Louis correspondent, was thrust into the ground two-thirds of its length.

In speaking of the tuning and receiving apparatus the electrical wizard stated:

"These vibrations reproduce sounds in receivers tuned to convey them to the listening ear. What this apparatus consists of or how it does work, I will not tell."

When Stubblefield returned to Murray and later became a recluse, his former friends advanced theories as to the cause of his failure.

"Those Easterners stole his invention when he went to Washington to demonstrate his invention," Dr. B.F. Berry told a correspondent recently.

His cousin, Vernon Stubblefield, believes the invention was stolen. The inventor, while in Washington, kept his valuables in a trunk. Perhaps this trunk was stolen. On the back of an old map of Washington, D.C., he wrote in a scrawling hand: "Will I ever see this trunk again? I believe these are honest Jews and they will do me right. N.B. Stubblefield, Feb. 13, 1912."

When he returned he advised his friends to withdraw their investments in his project.

"Damn rascals" was his only comment to friends who asked about eastern capitalists. The trunk was never seen again.

Perry Meloan, newspaper editor of Edmonton, Kentucky, who was an eye witness of the first public demonstration, declares that Stubblefield was inveigled into becoming a partner of the "Wireless Telephone Company of America" located at Broadway 11, New York.

Learning later that the company was not interested in perfecting the radio-telephone, but only in selling stock unscrupulously, Stubblefield came home. Of course the company kept the invention. This is the theory of Mr. Meloan.

That Stubblefield lacked executive ability was the reason advanced by President Wells, who, as an attorney, assisted him in obtaining his patents.

"Nathan Stubblefield was so suspicious of every investigator that he would not sign a contract for the commercializing of the venture," asserted Wells. "He was naturally suspicious of everyone. He thought he would be cheated."

Desire for greater personal glory and refusal to share fame with others undoubtedly prevented him from associating with capitalists who would finance his project. Only his son, a child of 14 years, knew anything about the principles of his invention. That boy, Bernard, now a grown man, is connected with the Westinghouse Electrical Corporation, which was the first concern to introduce the radio on a commercial basis. Did Bernard utilize his knowledge of his father's invention to aid the company in producing the radio?

In order to mystify and mislead onlookers and visitors, "crazy" Nat resorted to various tactics. On one occasion he invited a neighbor to listen while he broadcast. Handing the skeptical neighbor his receiving box, the shrewd genius commanded him to stand within an enclosure surrounded by a string tied on posts.

Taking his broadcasting set and talking as he walked Stubblefield moved methodically down the hill. Of course, the fellow-citizen thought that wires were buried beneath his feet. But the voice of Stubblefield came clear and strong.

"Can you hear me?" he asked. Then he whispered, whistled and played a harmonica.

When the inventor was out of sight, the neighbor stepped outside the string enclosure. Still the voice and music came as before. When the inventor returned, the neighbor was inside the string.

This desire to mislead the public may have stood in the way of those who would buy.

It is possible that some minor technical point in the invention escaped him, making it unmarketable. The contrivance, he admitted, was imperfect. Certainly he did not have sufficient electrical power to broadcast for great distances. Citizens of Murray say he was jealous of Marconi's success in developing wireless telegraphy and wished to keep from the world what he believed to be a greater contribution.

Domestic troubles may have interfered with his march to fame. Wedded to his science, he became morose, and indifferent to the needs of his family. He had nine children, six of whom are still living today.

"Because he worked eternally at his mysterious box", Mrs. Walter Stubblefield, wife of his brother, stated, "his wife and children left him."

Desertion by his friends and too much attention to his invention aggravated his mental peculiarity. Nevertheless, many of his neighbor Kentuckians still declare that he was never insane, but only "queer".

Strange tales are told by the superstitious Kentuckians about his last days. Living alone in his little hut, six miles north of Murray, near Almo, he was working on wireless plans.

Wireless lights appeared up in trees, on the end of steel rods, and along the woven wire fence, according to rumors spread by the people of Almo, Kentucky. Voices, coming apparently from the air, were frequently heard if trespassers are to be believed.

"Get your mule out of my cornfield", is a reported order that came by wireless to a neighbor when his animal had invaded the hermit's premises.

Becoming more and more suspicious of his fellowmen, Stubblefield became sick. He refused the aid of friends.

Robert McDermott found the dead body of Nathan Stubblefield in his little hut March 30, 1928. In an unmarked grave in Bowman's cemetery one and one-half miles north of Murray, his body lies alone. For two years no effort was made to give him glory.

On March 28, 1930, the little city will dedicate a modest marker on the campus of the college where Nathan B. Stubblefield first broadcast and received the human voice without wires.

"Inventor of Radio" is the title, which the citizens are justly placing on his monument.

Out there on the campus of Murray State Teacher's College the last speaker will be through. The marker will stand silent. But the visitor who reads the inscription on the monument will feel a thrill of wonder and perhaps sadness as a radio message from that other world says: "You and I will yet add luster to the Stubblefield name."

APPENDIX E.

The following is the text of an address by James L. Johnson, Executive Secretary of the Murray, Kentucky Chamber of Commerce to the Annual Convention of The Kentucky Broadcasters association in Louisville on May 18, 1961, reprinted here by permission of the author. Following the address, Charles Shuffet, representing the KBA, presented a plaque that officially recognized Nathan B. Stubblefield as the inventor of radio broadcast.

Nathan B. Stubblefield
"Father of Radio"

"Hello Rainey....Hello Rainey," these four words, highly insignificant in themselves, were the gateway that opened a fabulous new industry in the late 19th century and early 20th century. These were the first words ever broadcast by radio. These four words put you people in business.

When we think of today's marvels of scientific discovery, it is imperative that we think of highly trained research technicians working as a tightly-knit, well planned organization. We see on the one hand rack after rack of scientific apparatus...on the other the latest journals of technical information.

It is only natural that we tend to think of the invention of radio in the same light...but Nathan B. Stubblefield had none of these...yet he was all of them. He discovered what has probably become one of our most important scientific achievements, yet he learned not to care...He had an opportunity to gain vast wealth and worldwide recognition: but he died of starvation...He loved his family, but he lost them...He loved life, but he selfishly destroyed it...His mind was one of the world's greatest, and IT destroyed him.

This paragraph, a missile of utter confusion and frustration, literally depicts the life of Nathan B. Stubblefield. This confusion and frustration did not end with his death...but has plagued researchers since 1902 who have tried to piece together and verify the evidence that would rightly accord the honor due "The Father of Radio Broadcast." Nathan B. Stubblefield did more to muddle the clouded picture than anyone else did. In bitterness and hopelessness he methodically destroyed hundreds of prototypes that he had spent a miserable lifetime building. Basic radio broadcast equipment literally worth millions was stolen from his by ruthless promoters.

Perhaps, Nathan B. Stubblefield himself, realized better than anyone just how deeply embittered he had made himself, because he expressed himself eloquently in poems and prose, just a scant few years before he became embroiled in a bitter, world-wide scientific explosion. His expressions were so skillfully written that they bear out the claims of friends that Nathan was born a hundred years too soon.

Nathan B. Stubblefield was born in the summer of 1859, the son of William Jefferson and Victoria Bowman Stubblefield. Nathan attended the schools of Calloway County, and there his formal education ended. In his early youth he became enraptured with electrical experiments, that were interrupted just long enough for him to take a wife. Several children were born, four of who still live, and Nathan educated all of them himself. In spite of their constant exposure to electrical apparatus, only Bernard showed an interest in science. He became his father's chief, and only, assistant while just a child. Only Bernard was ever allowed to see the complete experimental equipment.

Early in life, Nathan acquired a reputation for being "peculiar", a tag he carried with him to his grave. Even as a boy he had few close friends...perhaps only Duncan Holt could be called his friend. Duncan Holt was a constant companion, because they both loved science and electricity. They spent hour after hour in the office of "The Calloway Times," edited by W.O. Wear. It wasn't that they thought so much of Mr. Wear...rather it was the only place they could find copies of the "Scientific American," the best technical magazine of the times.

Both Nathan and Duncan became obsessed with the experiments of Nikola Telsa, who was trying to send electrical impulses through Pikes Peak. Telsa has carried on extensive inquiry into the alternating current theory. The year was 1880, and Rudolf Hertz had already proved a startling fact. Hertz proved that electromagnetic waves, in transverse nature of vibrations, and that there susceptibility to reflection and polarization are in complete correspondence with waves of light and heat. Previously in 1864, Clerk Maxwell had figured out mathematically, that electromagnetic waves did exist.

Nathan studied these theories religiously, and became thoroughly familiar with the Hertzian Wave. At the time Nathan B. Stubblefield was 20 years old and Marconi...A SIX-YEAR-OLD CHILD...was bouncing around his father's Italian Villa in rompers!

Just two years later, in 1882, Nathan B. Stubblefield demonstrated to a Murray audience, that the earth current of electricity could and would drive a compass needle WILD...even though the compass and the electromagnetic generator were several yards apart. The experiment was a complete success...but did not impress the crowd. Mad...Nathan made up his mind to show them something that would make them sit up and take notice. His theory was that those same currents could carry the sound of the human voice and music...not only through the earth...but through the air as well.

At this time Marconi was only eight!

Nathan knew that he could make the machine to drive the sound through the air, but he didn't know how to build a machine to interpret and receive the signal. The problems involved were tremendous, the cost of materials was out of his reach ...but he knew that somehow, he would find the answer if he tried hard enough. Searching for money, he realized that Murray had no telephones...and he knew that Alexander Graham Bell had a patent on his phone...so he set about to build one of his own...It was a simple device...not handsome...but truly amazing. Nathan B. Stubblefield had built a telephone that worked without wires. This was the first radio in existence but Nate did not realize or grasp the significance of his achievement.

Marconi was 10 years old...and still living in Italy.

Nathan B. Stubblefield patented his Vibrating Telephone on February 21, 1888. Marconi was 14 at the time. Through the next four years, Stubblefield would be engaged in telephone building and installation. Letters in my possession attest to the fact that the Stubblefield telephone was far superior to the Bell invention.

In 1892 Nathan Stubblefield visited one of his few trusted friends, Rainey T. Wells, General Attorney for the Woodmen of the World. He asked Dr. Wells to visit him at his home...and Dr. Wells did. Little did he know that he had been selected to hear the first true radio voice broadcast in history. Affidavits made by Dr. Wells in 1892 tell of his participation in the experiment and of its complete success as far as he could tell. He was astounded, and immediately begged Stubblefield to patent his work, but Stubblefield steadfastly refused. He said it was far from PERFECT...and he wanted perfection. It was to be 15 long years before Dr. Wells and an eminent patent attorney Conn Linn, were able to persuade him that he should seek patent protection.

His experiment of 1892 soon leaked to the world of science, and a host of investors, promoters, swindlers, and scientists began to beat a path to the door of Stubblefield; they were met at the garden by Stubblefield and his shotgun. Offers from $50,000 to a half million dollars were made for a part of interest in the invention, but he refused all offers. For years he had worked with absolutely no help from anyone at all. He figured he had come this far alone...and would continue the same way.

In January of 1901 the Wireless Telephone Company of America was formed upon the basis of Stubblefield's invention of the Electric Battery, patented March 8, 1898 and upon his Wireless Telephone...before the radio patent was applied for.

In January of 1902 Stubblefield made his first public demonstration of his invention. This time newspapers were on hand for the event, and the St. Louis Post-Dispatch scored a tremendous scoop with their FULL-PAGE front coverage of the event. This newspaper as well as several thousand people saw and heard a convincing demonstration. Never again was he doubted. The world called for other demonstrations of this New World of magic...and Stubblefield obliged with demonstrations in Washington, in New York and Philadelphia. Still the patent had not been applied for.

During these demonstrations, the Wireless Telephone Company of America persuaded Stubblefield to join their ranks in the promotion of his invention. Stubblefield was reluctant, but he needed money to pursue and perfect Radio Broadcast. And now for the first time, I can reveal that Nathan B. Stubblfield traded all his interests, all his secrets, and all his equipment, for 500,000 shares of stock in this company. A company that had been formed upon his ideas, and had nothing except the price of their stationery invested. On May 14, 1903 Stubblefield discovered that his 500,000 shares of stock in a worthless company had been juggled so that the books read his shares as being 50,000. He protested and rather than let the information be revealed to the public, the company called it a typographical error and issued him a certificate for the original amount...which an official of the company, one Gerald M. Fennel, allegedly stole from Stubblefield. Of course the inventor protested vigorously...but he never regained his stock in the company...a stock that worthless in any case.

Stubblefield returned to New York to try and unravel the mystery that surrounded this company. It didn't take long. And now, I would like to read you a letter written by Nathan B. Stubblefield on June 19, 1902. This is the first time that the family has allowed the use of this material.

The letter was addressed:

Mr. Turner: Secretary of The Wireless Telephone Company of America Steven House, New York, New York

Dear Sir:

Mr. Gerald M. Fennel, the promoter of our company, had had letters from me, of which you have a copy. He has answered same, cleverly evading and practicing fraud of deception as usual...and there REMAINS NOTHING FOR ME TO DO BUT GO HOME. I regret very much that such had been the ending, and regret very much that my name is connected in any way with this concern. I shall take immediate steps when I reach home to turn on the lights that this fellow may not swindle the public as I have been. It becomes my duty, (as I am one of the directors) to see that this be a fair legitimate business or I am a party to the fraud that may be committed. I very respectfully decline having a thing to do with the business until it is in every way, put on an honorable basis; and put in the hands of men who will so conduct the same.

The letter referred to is of June 17, 1902 and is in nature, a complaint against the aforesaid promoters and should be seen by every member of the company. If you sir depend upon this man Fennel to put the matter before the company they will never know the facts as I have presented them. I therefore ask that you provide each of them with a copy of same, that they may have a chance to adjust this matter, after which time should they fail to act, then it remains to be clearly seen that they are parties to the swindle of me, out of my inventions, and the defrauding of the public. I shall notify each and all of them that you have such documents in your possession. To comply with my duty, since I HAVE SIGNED OVER EVERYTHING THAT I HAVE to this concern, I do today, with Mr. Wally Hood, your fellow associate clerk as witness, turn over to you all the property in my possession belonging to the company and depart for my Kentucky home with a feeling of gratitude for some New York people, who with me, have watched the steps of this man Fennel through many hours of uneasiness to me.

Signed: Nathan B. Stubblefield

Please note that still this invention has not been patented. Finally, on April 12, 1907 Stubblefield applied for a patent and was granted patent number #887357 on May 12, 1908. The same year Canada granted a patent for radio to Nathan B. Stubblefield. Obviously the patents were too late, because Collins Radio Company of Canada sold 75% of their rights to a process similar to Stubblefield's

Wireless Telephone Company of America for 71/2 million dollars. The company said the inventions were along the same lines.

In 1908, Nathan B. Stubblefield returned to his Calloway County home, which he had lost to creditors. He built himself a shack of tin, corn stalks and dirt on the floor. There for twenty years he brooded, silent and completely disillusioned. He refused to see or talk with anyone; he refused financial assistance or sympathy. On March 28, 1928 he was found dead in his shack, with only his half-starved cat for company. He obviously died of starvation on March 25, and his beloved cat had mutilated his eyes in search of water. Even in death, his dearest friend had turned against him.

The burial of Nathan B. Stubblefield did not settle the controversy that was aroused about this man. Researchers know that Nathan B. Stubblefield was truly a genius...but he himself has made it extremely difficult to prove.

In the minute of so that I have remaining, let's briefly review the accomplishments of Nathan B. Stubblefield:

Nathan B. Stubblefield invented and built his first vibrating telephone in 1882...patented in 1888.

He patented the electric storage battery, the basis of all his inventions, on March 8, 1898.

Rainey T. Wells heard the world's first broadcast early in 1892 in a private demonstration.

In 1890 he demonstrated the Stubblefield Telegraph System, a system that allowed you to dial the letter you wished to send. This is still not in existence today

On January 1, 1902 he demonstrated broadcast to Murray and the newspapers.

On March 20, 1902 the first ship to shore transmission was made from the steamer Bartholdi to Stubblefield on the banks of the Potomac River.

On May 30, 1902, a group of the world's leading scientists saw a public demonstration of the broadcast system at Belmont Park, Philadelphia.

On April 12, 1902 Stubblefield applied for patents and was granted Patent Number 887357 on May 12, 1908 for the Wireless Telephone.

March 25, 1928, Stubblefield died of starvation.

In 1930 the New York Supreme Court ruled that heirs of Stubblefield had proved every detail in their claim for patent rights, but that the Statute of Limitations made their claims void as to royalties

On March 28, 1930, grateful citizens of Murray erected a monument to Stubblefield on the campus of Murray State College. This was the full price...The WORLD PAID FOR RADIO.

REFERENCES

Most of the background material for this book came from the Nathan B. Stubblefield Papers, stored at the Pogue Library of Murray State University, in Murray, Kentucky. This archive consists of original documents, photographs, clippings, articles, and unpublished manuscripts. There are also files of correspondence from L.J. Hortin, James Johnson, Vernon Stubblefield, the Murray Chamber of Commerce and others, relative to Nathan Stubblefield. There is a smaller but impressive collection of Stubblefield memorabilia at the Wrather Museum, also on Murray State University campus. I urge anyone who wants to know more about Nathan to spend a few days browsing through these primary sources.

I pieced together the narrative of Nathan's life from these archives, from the L.J. Hortin article reprinted as Appendix D., and from two dissertations, which are available through University Microfilms:

The Role of the Independent Inventor in the Early Development of Electrical Technology. David Hendrick Miller. University of Missouri - Columbia, 1971.

The Contribution of Nathan B. Stubblefield to the Invention of Wireless Voice Communications. Thomas Olin Morgan. Florida State University, 1971.

There are many sources for information about wireless inventors and inventions in the period when Nathan worked. Three generally available books are:

Syntony and Spark: The Origins of Radio. Hugh G.J. Aitken. John Wiley and Sons, 1976.

History of Radio to 1926. Gleason L. Archer. Arno Press, 1971.

A History of Wireless Telegraphy. J.J. Fahie. Arno Press, 1971.

You can find a good concise summary of the period, with a short section about Nathan, in this article:

"A Technological Survey of Broadcasting's 'Pre-History,' 1876-1920." Elliot N. Sivowitch. *Journal of Broadcasting*, Winter 1970-1971.

All copies of patents came from the US Patent Office. Patent files, including the details of the patent interference case involving Granville Woods, Lucius Phelps, and Thomas Edison, are kept at the National Archives. The background material on the wireless inventions of Samuel F.B. Morse and Alexander Graham Bell came mostly from original notebooks and other documents in the Manuscript Division, the Library of Congress. The information about RCA in Chapter 9 came from correspondence files in the George Howard Clark Papers, National Museum of American History, at the Smithsonian Institution.

HERE LIE THE EARTHLY
REMAINS OF
NATHAN B.
STUBBLEFIELD
NOV. 22, 1860
MAR. 28, 1928
INVENTOR OF WIRELESS
TELEPHONY OR RADIO

R.I.P

IF YOU WOULD LIKE ADDITIONAL COPIES OF THIS BOOK, PLEASE VISIT THE *ALL ABOUT WIRELESS* WEB PAGE FOR ALL THE DETAILS.

YOU CAN FIND US AT

WWW.NATHANSTUBBLEFIELD.COM

THANK YOU.